LONDON MATHEMATICAL SOCIETY LECTURE NOTE SERIES

Managing Editor: Professor J.W.S. Cassels, Department of Pure Mathematics and Mathematical Statistics, University of Cambridge, 16 Mill Lane, Cambridge CB2 1SB, England

The books in the series listed below are available from booksellers, or, in case of difficulty, from Cambridge University Press.

17 Differential germs and catastrophes, Th. BROCKER & L. LANDER
34 Representation theory of Lie groups, M.F. ATIYAH *et al*
36 Homological group theory, C.T.C. WALL (ed)
39 Affine sets and affine groups, D.G. NORTHCOTT
40 Introduction to H_p spaces, P.J. KOOSIS
43 Graphs, codes and designs, P.J. CAMERON & J.H. VAN LINT
45 Recursion theory: its generalisations and applications, F.R. DRAKE & S.S. WAINER (eds)
46 p-adic analysis: a short course on recent work, N. KOBLITZ
49 Finite geometries and designs, P. CAMERON, J.W.P. HIRSCHFELD & D.R. HUGHES (eds)
50 Commutator calculus and groups of homotopy classes, H.J. BAUES
51 Synthetic differential geometry, A. KOCK
54 Markov processes and related problems of analysis, E.B. DYNKIN
57 Techniques of geometric topology, R.A. FENN
58 Singularities of smooth functions and maps, J.A. MARTINET
59 Applicable differential geometry, M. CRAMPIN & F.A.E. PIRANI
60 Integrable systems, S.P. NOVIKOV *et al*
62 Economics for mathematicians, J.W.S. CASSELS
65 Several complex variables and complex manifolds I, M.J. FIELD
66 Several complex variables and complex manifolds II, M.J. FIELD
68 Complex algebraic surfaces, A. BEAUVILLE
69 Representation theory, I.M. GELFAND *et al*
74 Symmetric designs: an algebraic approach, E.S. LANDER
76 Spectral theory of linear differential operators and comparison algebras, H.O. CORDES
77 Isolated singular points on complete intersections, E.J.N. LOOIJENGA
78 A primer on Riemann surfaces, A.F. BEARDON
79 Probability, statistics and analysis, J.F.C. KINGMAN & G.E.H. REUTER (eds)
80 Introduction to the representation theory of compact and locally compact groups, A. ROBERT
81 Skew fields, P.K. DRAXL
82 Surveys in combinatorics, E.K. LLOYD (ed)
83 Homogeneous structures on Riemannian manifolds, F. TRICERRI & L. VANHECKE
85 Solitons, P.G. DRAZIN
86 Topological topics, I.M. JAMES (ed)
87 Surveys in set theory, A.R.D. MATHIAS (ed)
88 FPF ring theory, C. FAITH & S. PAGE
89 An F-space sampler, N.J. KALTON, N.T. PECK & J.W. ROBERTS
90 Polytopes and symmetry, S.A. ROBERTSON
91 Classgroups of group rings, M.J. TAYLOR
92 Representation of rings over skew fields, A.H. SCHOFIELD
93 Aspects of topology, I.M. JAMES & E.H. KRONHEIMER (eds)
94 Representations of general linear groups, G.D. JAMES
95 Low-dimensional topology 1982, R.A. FENN (ed)
96 Diophantine equations over function fields, R.C. MASON
97 Varieties of constructive mathematics, D.S. BRIDGES & F. RICHMAN
98 Localization in Noetherian rings, A.V. JATEGAONKAR
99 Methods of differential geometry in algebraic topology, M. KAROUBI & C. LERUSTE
100 Stopping time techniques for analysts and probabilists, L. EGGHE

101 Groups and geometry, ROGER C. LYNDON
103 Surveys in combinatorics 1985, I. ANDERSON (ed)
104 Elliptic structures on 3-manifolds, C.B. THOMAS
105 A local spectral theory for closed operators, I. ERDELYI & WANG SHENGWANG
106 Syzygies, E.G. EVANS & P. GRIFFITH
107 Compactification of Siegel moduli schemes, C-L. CHAI
108 Some topics in graph theory, H.P. YAP
109 Diophantine Analysis, J. LOXTON & A. VAN DER POORTEN (eds)
110 An introduction to surreal numbers, H. GONSHOR
111 Analytical and geometric aspects of hyperbolic space, D.B.A.EPSTEIN (ed)
112 Low-dimensional topology and Kleinian groups, D.B.A. EPSTEIN (ed)
113 Lectures on the asymptotic theory of ideals, D. REES
114 Lectures on Bochner-Riesz means, K.M. DAVIS & Y-C. CHANG
115 An introduction to independence for analysts, H.G. DALES & W.H. WOODIN
116 Representations of algebras, P.J. WEBB (ed)
117 Homotopy theory, E. REES & J.D.S. JONES (eds)
118 Skew linear groups, M. SHIRVANI & B. WEHRFRITZ
119 Triangulated categories in the representation theory of finite-dimensional algebras, D. HAPPEL
120 Lectures on Fermat varieties, T. SHIODA
121 Proceedings of *Groups - St Andrews 1985*, E. ROBERTSON & C. CAMPBELL (eds)
122 Non-classical continuum mechanics, R.J. KNOPS & A.A. LACEY (eds)
123 Surveys in combinatorics 1987, C. WHITEHEAD (ed)
124 Lie groupoids and Lie algebroids in differential geometry, K. MACKENZIE
125 Commutator theory for congruence modular varieties, R. FREESE & R. MCKENZIE
126 Van der Corput's method for exponential sums, S.W. GRAHAM & G. KOLESNIK
127 New directions in dynamical systems, T.J. BEDFORD & J.W. SWIFT (eds)
128 Descriptive set theory and the structure of sets of uniqueness, A.S. KECHRIS & A. LOUVEAU
129 The subgroup structure of the finite classical groups, P.B. KLEIDMAN & M.W.LIEBECK
130 Model theory and modules, M. PREST
131 Algebraic, extremal & metric combinatorics, M-M. DEZA, P. FRANKL & I.G. ROSENBERG (eds)
132 Whitehead groups of finite groups, ROBERT OLIVER
133 Linear algebraic monoids, MOHAN S. PUTCHA
134 Number theory and dynamical systems, M. DODSON & J. VICKERS (eds)
135 Operator algebras and applications, 1, D. EVANS & M. TAKESAKI (eds)
136 Operator algebras and applications, 2, D. EVANS & M. TAKESAKI (eds)
137 Analysis at Urbana, I, E. BERKSON, T. PECK, & J. UHL (eds)
138 Analysis at Urbana, II, E. BERKSON, T. PECK, & J. UHL (eds)
139 Advances in homotopy theory, S. SALAMON, B. STEER & W. SUTHERLAND (eds)
140 Geometric aspects of Banach spaces, E. M. PEINADOR and A. RODES (eds)
141 Surveys in combinatorics 1989, J. SIEMONS (ed)
142 The geometry of jet bundles, D.J. SAUNDERS
143 The ergodic theory of discrete groups, PETER J. NICHOLLS

London Mathematical Society Lecture Note Series. 139

Advances in Homotopy Theory

Proceedings of a conference in honour of the 60th birthday of I.M. James

Edited by

S.M. Salamon
B. Steer
W.A. Sutherland

Mathematical Institute, Oxford University

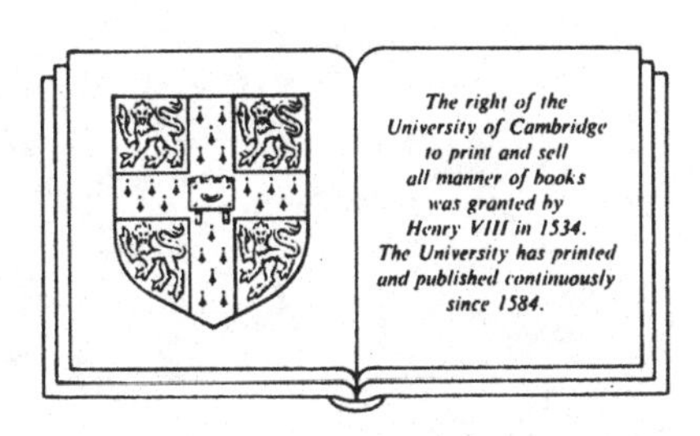

CAMBRIDGE UNIVERSITY PRESS
Cambridge
New York Port Chester Melbourne Sydney

Published by the Press Syndicate of the University of Cambridge
The Pitt Building, Trumpington Street, Cambridge CB2 1RP
40 West 20th Street, New York, NY 10011, USA
10, Stamford Road, Oakleigh, Melbourne 3166, Australia

First published 1989

Library of Congress cataloging in publication data available

British Library cataloguing in publication data available

ISBN 0 521 37907 5

In honour of Ioan James
on the occasion of his sixtieth birthday

Contents

Preface

In 1988 Ioan James celebrated his sixtieth birthday. To mark the occasion, and to recognize his contribution to mathematics, the Fifth Oxford Topology Symposium took the form of a conference on Homotopy Theory at the Palazzone in Cortona, Italy, held with the assistance of the Scuola Normale Superiore, Pisa. The conference enjoyed financial support from the London Mathematical Society, Cambridge University Press, and Pergamon Press.

This volume is intended both to be a record of proceedings, and to give some insight into the future development of Homotopy Theory. The articles range from almost verbatim accounts of talks describing current work to more detailed papers, which in a few cases discuss material not presented at the conference. The Editors thank all the contributors very warmly, as well as the staff of Cambridge University Press.

On behalf of the Organizing Committee for the conference, we acknowledge help, not only from the sponsors listed above, but also from those institutions, including the Royal Society and the National Science Foundation, that helped with the fares of individual participants. We are grateful to the Scuola Normale for allowing us to hold the Symposium in such beautiful surroundings, and specially thank all those in Pisa whose help was so essential for its successful organization. In addition, we record our warm appreciation for the efficient and courteous secretarial services of Silvana Boscherini, who stood in at short notice. Finally, we thank all participants, and especially Ioan James, for making the conference so profitable and enjoyable.

Simon Salamon
Brian Steer
Wilson Sutherland

List of participants at conference

*indicates speaker and contributor to this volume

M.G. Barratt	Northwestern
E.H. Brown	Brandeis
S. Buoncristiano	Rome II
*G. Carlsson	Princeton
*M.C. Crabb	Aberdeen
*F.R. Cohen	Kentucky
J.M. Cohen	Maryland and Bari
R.L. Cohen	Stanford
E. Devinatz	Chicago
A. Dold	Heidelberg
*W.G. Dwyer	Notre Dame
E.M. Friedlander	Northwestern
L. Hodgkin	King's College, London
*M.J. Hopkins	Chicago
*J.R. Hubbuck	Aberdeen
N. Iwase	Kyushu and Aberdeen
I.M. James	Oxford
J.D.S. Jones	Warwick
R. Kane	London, Ontario
*N.J. Kuhn	Virginia
*J. Lannes	École Polytechnique
M. Mahowald	Northwestern
*H.R. Miller	M.I.T.
G. Mislin	Ohio and Zurich
*S.A. Mitchell	Washington, Seattle
J. Neisendorfer	Rochester
F.P. Peterson	M.I.T.
S. Priddy	Northwestern
D. Puppe	Heidelberg
*E. Rees	Edinburgh

*C.A. Robinson	Warwick
S.M. Salamon	Oxford
*G.B. Segal	Oxford
*J.D. Stasheff	North Carolina
B. Steer	Oxford
C.T. Stretch	Coleraine
W.A. Sutherland	Oxford
E. Thomas	Berkeley
R. Thompson	Northwestern
G. Walker	Manchester
C. Wilkerson	Wayne State and Cornell
*R.M.W. Wood	Manchester

Addresses of Contributors

G. Carlsson
Department of Mathematics,
Fine Hall, Washington Road,
Princeton, NJ 08544-1000,
U.S.A.

F.R. Cohen
Department of Mathematics,
University of Kentucky,
Lexington, KY 40506-0027,
U.S.A.

M.C. Crabb
Department of Mathematics,
The Edward Wright Building,
Dunbar Street,
Aberdeen, AB9 2TY,
U.K.

Duan H.
Institute of Systems Science,
Academia Sinica,
Beijing,
China.

W.G. Dwyer
Department of Mathematics,
University of Notre Dame,
P.O. Box 398,
Notre Dame, IN 46556,
U.S.A.

J.H. Gunawardena
Trinity College,
Cambridge, CB2 1TQ,
U.K.

J.R. Harper
Department of Mathematics,
University of Rochester,
Rochester, NY 14627,
U.S.A.

M.J. Hopkins
Department of Mathematics,
University of Chicago,
Chicago, IL 60637,
U.S.A.

J.R. Hubbuck
Department of Mathematics,
The Edward Wright Building,
Dunbar Street,
Aberdeen, AB9 2TY,
U.K.

D.M. Kan
Department of Mathematics,
Massachusetts Institute of Technology,
Cambridge, MA 02139,
U.S.A.

N.J. Kuhn
Department of Mathematics,
University of Virginia,
Charlottesville, VA 22903,
U.S.A.

J. Lannes
Centre de Mathématiques,
École Polytechnique,
91128 Palaiseau Cedex,
France.

H.R. Miller
Department of Mathematics,
Massachusetts Institute of Technology,
Cambridge, MA 02139,
U.S.A.

S.A. Mitchell
Department of Mathematics,
University of Washington,
Seattle, WA 98195,
U.S.A.

E. Rees
Department of Mathematics,
University of Edinburgh,
King's Buildings,
Mayfield Road,
Edinburgh, EH9 3JZ,
U.K.

C.A. Robinson
Mathematics Institute,
University of Warwick,
Coventry, CV4 7AL,
U.K.

G.B. Segal
Mathematical Institute,
24–29 St. Giles',
Oxford, OX1 3LB,
U.K.

J.D. Stasheff
Department of Mathematics,
Univerity of North Carolina,
Chapel Hill, NC 27599–3455,
U.S.A.

R.M.W. Wood
Department of Mathematics,
The University,
Manchester, M13 9PL,
U.K.

S. Zarati
Faculté des Sciences-Mathématiques,
Université de Tunis,
1060 Tunis,
Tunisia.

Publications of I. M. James

1. Note on factor spaces. *J. London Math. Soc.* 28 (1953), 278–285.

2. (with J.H.C. Whitehead) Note on fibre spaces. *Proc. London Math. Soc. (3)*, 4 (1954), 129–137.

3. (with J.H.C. Whitehead) The homotopy theory of sphere bundles over spheres. I. *Proc. London Math. Soc. (3)*, 4 (1954), 196–218.

4. (with J.H.C. Whitehead) The homotopy theory of sphere bundles over spheres. II. *Proc. London Math. Soc. (3)*, 5 (1955), 148–166.

5. On the iterated suspension. *Quart. J. Math. Oxford Ser. (2)*, 5 (1954), 1–10.

6. On the homotopy groups of certain pairs and triads. *Quart. J. Math. Oxford Ser. (2)*, 5 (1954), 260–270.

7. Reduced product spaces. *Ann. of Math. (2)*, 62 (1955), 170–197.

8. On the suspension triad. *Ann. of Math. (2)*, 63 (1956), 191–247.

9. The suspension triad of a sphere. *Ann. of Math. (2)*, 63 (1956), 407–429.

10. Symmetric functions of several variables, whose range and domain is a sphere. *Bol. Soc. Mat. Mexicana (2)*, 1 (1956), 85–88.

11. On the suspension sequence. *Ann. of Math. (2)*, 65 (1957), 74–107.

12. Note on cup-products. *Proc. Amer. Math. Soc.* 8 (1957) 374–383.

13. Multiplication on spheres. I. *Proc. Amer. Math. Soc.* 8 (1957), 192–196.

14. Multiplication on spheres. II. *Trans. Amer. Math. Soc.* 84 (1957), 545–558.

15. On spaces with a multiplication. *Pacific J. Math.* 7 (1957), 1083–1100.

16. Commutative products on spheres. *Proc. Cambridge Philos. Soc.* 53 (1957), 63–68.

17. Whitehead products and vector fields on spheres. *Proc. Cambridge Philos. Soc.* 53 (1957), 817–820.

18. On the homotopy groups of spheres. *Proc. International Symposium on Algebraic Topology, pp. 222–224, Universidad Autónoma de Mexico and UNESCO*, 1958.

19. Filtration of the homotopy groups of spheres. *Quart. J. Math. Oxford Ser. (2)*, 9 (1958), 301–309.

20. Embeddings of real projective spaces. *Proc. Cambridge Philos. Soc.* 54 (1958), 555–557.

21. (with J.H.C. Whitehead) Homology with zero coefficients. *Quart. J. Math. Oxford Ser. (2)*, 9 (1958), 317–320.

22. The intrinsic join: a study of the homotopy groups of Stiefel manifolds, *Proc. London Math. Soc. (3)*, 8 (1958), 507–535.

23. Cross-sections of Stiefel manifolds. *Proc. London Math. Soc. (3)*, 8 (1958), 536–547.

24. Spaces associated with Stiefel manifolds. *Proc. London Math. Soc. (3)*, 9 (1959), 115–140.

25. Some embeddings of projective spaces. *Proc. Cambridge. Philos. Soc.* 55 (1959), 294–298.

26. Products on spheres. *Mathematika* 6 (1959), 1–13.

27. On Lie groups and their homotopy groups. *Proc. Cambridge Philos. Soc.* 55 (1959), 244–247.

28. (with E. Thomas) Which Lie groups are homotopy-abelian? *Proc. Nat. Acad. Sci. U.S.A.* 45 (1959), 737–740.

29. (with M.G. Barratt and N. Stein) Whitehead products and projective spaces. *J. Math. Mech.* 9 (1960), 813–819.

30. (with S. Araki and E. Thomas) Homotopy-abelian Lie groups. *Bull. Amer. Math. Soc.* 66 (1960), 324–326.

31. On H-spaces and their homotopy groups. *Quart. J. Math. Oxford Ser. (2)*, 11 (1960), 161–179.

32. On sphere-bundles over spheres. *Comment. Math. Helv.* 35 (1961), 126–135.

33. The transgression and Hopf invariant of a fibration. *Proc. London Math. Soc. (3)*, 11 (1961), 588–600.

34. Suspension of transgression. *Fund. Math.* 50 (1962), 501–507.

35. (with E. Thomas) Homotopy-commutativity in rotation groups. *Topology* 1 (1962), 121–124.

36. (with E. Thomas) Homotopy-abelian topological groups. *Topology* 1 (1962), 237–240.

37. (with E. Thomas) On homotopy-commutativity. *Ann. of Math. (2)*, 76 (1962), 9–17.

38. (edited) The Collected Works of J.H.C. Whitehead. I–IV. Pergamon Press, 1963.

39. The space of bundle maps. *Topology* 2 (1963), 45–59.

40. On the immersion problem for real projective spaces. *Bull. Amer. Math. Soc.* 69 (1963), 231–238.

41. (with E. Thomas, H. Toda and G.W. Whitehead) On the symmetric square of a sphere. *J. Math. Mech.* 12 (1963), 771–776.

42. Quasigroups and topology. *Math. Z.* 84 (1964), 329–342.

43. (with E. Thomas) An approach to the enumeration problem for non-stable vector bundles. *J. Math. Mech.* 14 (1965), 485–506.

44. (with E. Thomas) Note on the classification of cross-sections. *Topology* 4 (1966), 351–359.

45. (with E. Thomas) On the enumeration of cross-sections. *Topology* 5 (1966), 95–114.

46. A relation between Postnikov classes. *Quart. J. Math. Oxford Ser. (2)*, 17 (1966), 269–280.

47. On the homotopy theory of the classical groups. *An. Acad. Brasil. Ci.* (1967), 39–44.

48. (with E. Thomas) Submersions and immersions of manifolds. *Invent. Math.* 2 (1967), 171–177.

49. On homotopy-commutativity. *Topology* 6 (1967), 405–410.

50. Bundles with special structure. I. *Ann. of Math. (2)*, 89 (1969), 359–390.

51. On sphere bundles. I. *Bull. Amer. Math. Soc.* 75 (1969), 617–621.

52. On sphere bundles. II. *Bull. London Math. Soc.* 1 (1969), 323–328.

53. On fibre bundles and their homotopy groups. *J. Math. Kyoto Univ.* 9 (1969), 5–24.

54. On the Bott suspension. *J. Math. Kyoto Univ.* 9 (1969), 161–188.

55. Note on Stiefel manifolds. I. *Bull. London Math. Soc.* 2 (1970), 199–203.

56. Overhomotopy theory. *Proc. conf. on Homological Algebra and Algebraic Topology, INdAM, Rome, 1969. Symposia Mathematica 4, pp. 219–229, Academic Press*, 1970.

57. On the decomposability of fibre spaces. *The Steenrod Algebra and its Applications (Proc. conf. to celebrate N.E. Steenrod's sixtieth birthday, Columbus, Ohio, 1970). Lecture Notes in Math. 168, pp. 125–134, Springer*, 1970

58. Bibliography on H-spaces. *Proc. conf. on H-spaces, Neuchatel, 1970. Lecture Notes in Math. 196, pp. 137–156, Springer*, 1971.

59. On the homotopy-symmetry of sphere bundles. *Proc. Cambridge Philos. Soc.* 69 (1971), 291–294.

60. On the homotopy type of Stiefel manifolds. *Proc. Amer. Math. Soc.* 29 (1971), 151–158.

61. Ex-homotopy theory. I. *Illinois J. Math.* 15 (1971), 324–337.

62. Euclidean models of projective spaces. *Bull. London Math. Soc.* 3 (1971), 257–276.

63. On the maps of one fibre spaces into another. *Compositio Math.* 23 (1971), 317–328.

64. Note on Stiefel manifolds. II. *J. London Math. Soc. (2)*, 4 (1971), 109–117.

65. On sphere-bundles with certain properties. *Quart. J. Math. Oxford Ser. (2)*, 22 (1971), 353–370.

66. Products between homotopy groups. *Compositio Math.* 23 (1971), 329–345.

67. (with J. Adem and S. Gitler) On axial maps of a certain type. *Bol. Soc. Mat. Mexicana (2)*, 17 (1972), 59–62.

68. Two problems studied by Heinz Hopf. *Lectures on algebraic and differential topology (Second Latin American School in Math., Mexico City, 1971). Lecture Notes in Math. 279, pp. 134–174, Springer*, 1972.

69. (with D.W. Anderson) Bundles with special structure. II. *Proc. London Math. Soc. (3)*, 24 (1972), 324–330.

70. (with D.W. Anderson) Bundles with special structure. III. *Proc. London Math. Soc. (3)*, 24 (1972), 331–347.

71. (with W.A. Sutherland) On stunted real projective spaces. *Quart. J. Math. Oxford Ser. (2)*, 25 (1974), 101–112.

72. Which fibre spaces are decomposable? *Indag. Math.* 37 (1975), 385–390.

73. Homotopy-equivariance. *Comment. Math. Helv.* 50 (1975), 521–533.

74. The Topology of Stiefel Manifolds. London Math. Soc. Lect. Note Ser. 24, Cambridge University Press, 1976.

75. Relative Stiefel manifolds. *J. London Math. Soc. (2)*, 13 (1976), 331–335.

76. Alternative homotopy theories. *Enseignement Math. (2)*, 23 (1977), 221–237.

77. On fibre spaces and nilpotency. *Math. Proc. Cambridge Philos. Soc.*, 84 (1978), 57–60.

78. The works of Daniel Quillen. *Proc. International Congress of Mathematicians (Helsinki, 1978), pp. 65–67, Acad. Sci. Fennica*, 1980.

79. (with G.B. Segal) On equivariant homotopy type. *Topology* 17 (1978), 267–272.

80. On category, in the sense of Lusternik-Schnirelmann. *Topology* 17 (1978), 331–348.

81. On fibre spaces and nilpotency. II. *Math. Proc. Cambridge Philos. Soc.* 86 (1979), 215–217.

82. (with G.B. Segal) On equivariant homotopy theory. *Topology Symposium, Siegen 1979. Lecture Notes in Math. 788, pp. 316–330, Springer*, 1980.

83. On category and sectional category. *Proc. 1979 Topology Conference. Brazilian Math. Soc., APUCI*, 1980.

84. (edited) Topological Topics. London Math. Soc. Lect. Note Ser. 86, Cambridge University Press, 1983.

85. General Topology and Homotopy Theory, Springer, 1984.

86. General topology over a base. *Aspects of Topology. London Math. Soc. Lect. Note Ser. 93, pp. 141–174, Cambridge University Press*, 1985.

87. (edited, with E.R. Kronheimer) Aspects of Topology. London Math. Soc. Lect. Note Ser. 93, Cambridge University Press, 1985.

88. Uniform spaces over a base. *J. London Math. Soc. (2)*, 32 (1985), 328–336.

89. Spaces. *Bull. London Math. Soc.* 18 (1986), 529–559.

90. Topological and Uniform Spaces. Springer, 1987.

91. Fibrewise Topology. Cambridge University Press, 1988.

Homotopy fixed points in the algebraic K-theory of certain infinite discrete groups

Gunnar Carlsson*

1 Introduction

Since their introduction 15 years ago, Quillen's algebraic K-groups of rings have remained fairly difficult to compute in most cases of interest. This paper is a brief sketch of an attempt to remedy this situation in the case of group rings of certain discrete groups. Group rings are of particular interest since the algebraic K-groups are in this case related to the geometry of manifolds with the given group as fundamental group. Specifically, the algebraic K-groups of $\mathbb{Z}(\pi_1 M)$ are involved in the description of the space of self-homeomorphisms of M.

Recall that the zero-th space of Quillen's K-theory spectrum associated to a ring A is the space $BGL^+(A)$ [7]. One obvious map into the space $BGL^+(\mathbb{Z}\Gamma)$ arises from the existence of a tensor product map $BGL_n(A) \times BGL_m(B) \xrightarrow{\text{“}\otimes\text{”}} BGL_{n \cdot m}(A \otimes_C B)$, where A and B are C-algebras, as follows. We note that the group Γ is contained in the group $GL_1(\mathbb{Z}\Gamma)$, as one-by-one matrices with entry the given element of Γ. Thus, we have a map $\Gamma \times GL_n\mathbb{Z} \xrightarrow{\text{“}\otimes\text{”}} GL_n\mathbb{Z}\Gamma$. Applying the classifying space functor and passing to the limit over n we obtain a map $B\Gamma \times BGL^+\mathbb{Z} \to BGL^+(\mathbb{Z}\Gamma)$. After suitable interpretation, one finds that this map can be delooped so as to obtain a map of spectra $B\Gamma_+ \wedge \underset{\sim}{K}\mathbb{Z} \xrightarrow{\alpha} \underset{\sim}{K}\mathbb{Z}\Gamma$, where $\underset{\sim}{K}$ denotes the K-theory spectrum, and α is called the assembly map. In many cases, one conjectures that the map is a split injection onto a wedge summand, or perhaps that it is an equivalence. Much work has been done in this direction.

(a) Waldhausen [10] showed the map to be an equivalence in many cases, and analysed the failure of the map to be an equivalence in many others. He studied groups constructed from infinite cyclic ones by amalgamated sum and Laurent extension procedures.

(b) Quinn showed that if Γ is the fundamental group of a flat manifold, then the assembly map is an equivalence after rationalization [8].

*Supported in part by National Science Foundation grant DMS–8704668.A01

(c) The L-theory analogue of the conjecture that α is injective on a wedge summand is a strong form of the Novikov conjecture. The actual Novikov conjecture is that α is an injection after rationalization. The K-theory version has been proved for groups with finitely generated homology by Bökstedt, Hsiang, and Madsen.

(d) The A-theory analogue of α has been studied with spectacular success by Farrell and Jones [5]. A-theory means Waldhausen's K-theory of the spaces applied to $B\Gamma$. Farrell and Jones completely describe $A(B\Gamma)$, when Γ is the fundamental group of a negatively curved manifold. A corollary of their work is that the ordinary K-theory assembly map is an equivalence after rationalization for these groups.

2 A homotopy-theoretic approach

We'll describe an approach to this problem fundamentally different from those used in the above pieces of work. To motivate the procedures, we'll consider first the case of the complex group ring of a finite group. Here, Atiyah [1] and Atiyah-Segal [2] observed that the K-theory spectrum could be viewed as the fixed point spectrum of the group G acting on a G-equivariant spectrum having the homotopy type of $\underset{\sim}{K}\mathbf{C}$, which we shall call $K_G(\mathbf{C})$. The zero-th space of $K_G(\mathbf{C})$ is the space BU, equipped with an action arising from all complex representations of G. Now $BU^G \cong \coprod_\rho BU$, where ρ ranges over all virtual representations of G. Atiyah [1] went on to show that if we consider the map $BU^G \to BU^{hG}$, where BU^{hG} denotes the homotopy fixed point set, we often detect the entire space BU^G, and hence the K-theory spectrum of $\mathbf{C}[G]$. See [3] for a discussion of homotopy fixed point sets. Specifically, if G is a p-group Atiyah showed that the map $BU^G \to BU^{hG}$ is an equivalence after p-adic completion. This suggests that we attempt to mimic this procedure, even in the case of infinite groups. We arrive immediately at a problem, since a key point in the Atiyah-Segal program was that G acted on *finite* matrices. When we naively generalize this part of the program, we find that $\mathbf{Z}\Gamma$ is indeed realizable as the fixed point subring of an action of Γ on a ring of matrices, but that the matrices are necessarily infinite. It is easy to show that when one considers the ring of all infinite matrices with entries in $\mathbf{Z}$, whose rows and columns are finite, the associated K-theory spectrum is contractible, and hence so is its homotopy fixed point set. Further, the natural guess for a spectral sequence converging to $K_*\mathbf{Z}\Gamma$ would have $E_2^{p,q}$-term $H^{-p}(\Gamma; K_q\mathbf{Z})$, and many negative-dimensional groups would arise. The second observation shows that this kind of naive approach will not lead very far.

Our new approach can be summarized as follows. Roughly speaking, instead of considering all finite matrices, one should consider an appropriately chosen subring. In the case of $\Gamma = \mathbf{Z}$, one will consider a subring of the ring of all endomorphisms of the free $\mathbf{Z}$-module with basis $\{e_n\}_{n=-\infty}^{\infty}$. The subring in question consists of all those matrices M for which there is a number $N(M)$ so that $M_{ij} = 0$ if $|i-j| > N(M)$. Here, M_{ij} denotes the matrix entry associated to i and j. This subring will be called the ring of *bounded* endomorphisms of the given module, and will be denoted by $\mathcal{B}$. Now, Γ acts on the basis $\{e_n\}$ by $\sigma e_n = e_{n+1}$, where σ is a generator for Γ. This action gives a conjugation action on $\mathcal{B}$, and the fixed point subring is $\mathbf{Z}(\Gamma)$. The important point is that the K-theory of $\mathcal{B}$ is not trivial, but in fact $K_i(\mathcal{B}) = K_{i-1}(\mathbf{Z})$, and $\underset{\sim}{K}\mathcal{B} \cong \Sigma \underset{\sim}{K}\mathbf{Z}$. When we take the homotopy fixed point set $\underset{\sim}{K}\mathcal{B}^{h\Gamma}$, we find that $\underset{\sim}{K}\mathcal{B}^{h\Gamma} \cong \underset{\sim}{K}\mathbf{Z} \vee \Sigma \underset{\sim}{K}\mathbf{Z}$, which is known to be the correct answer for $\underset{\sim}{K}\mathbf{Z}(\Gamma)$ from the localization sequence and homotopy axiom for algebraic K-theory [7]. This suggests very strongly that we are on the right track, and should attempt to find a suitable replacement for $\mathcal{B}$ for a more general class of groups. The suitable framework is the bounded algebraic K-theory of E. Pedersen and C. Weibel [6]. They associate to any metric space X and ring R a spectrum $\underset{\sim}{K}(R;X)$, the K-theory of R with labels in X. This is done as follows. To R and X, we first associate a category $\mathcal{C}_X(R)$, whose objects are free R-modules equipped with a basis $B = \{e_\alpha\}_{\alpha \in A}$ and a "labelling" function $\varphi: B \to X$. The modules may be infinitely generated. A morphism from $\{M, B_M, \varphi_M\}$ to $\{N, B_N, \varphi_N\}$ is an R-linear isomorphism f which is bounded in the following sense. Given a linear transformation $T: M \to N$, let $\{T_{\alpha\beta}\}_{\alpha \in B_M, \beta \in B_N}$ be the matrix of T relative to the bases B_M and B_N. Then f is said to be bounded if there is a number L so that $f_{\alpha\beta} = 0 = (f^{-1})_{\beta\alpha}$ if $d(\varphi_M(\alpha), \varphi_N(\beta)) > L$. The category $\mathcal{C}_X(R)$ is symmetric monoidal, and so is its "idempotent completion" $\widehat{\mathcal{C}}_X(R)$. To any symmetric monoidal category $\mathcal{A}$ one associates a spectrum, as in [9], called $\mathrm{Spt}\mathcal{A}$. Now $\underset{\sim}{K}(R, X)$ is defined to be $\mathrm{Spt}(\widehat{\mathcal{C}}_X(R))$.

Pedersen and Weibel proceed to prove several results about their construction. For instance, if E^n denotes $\mathbf{R}^n$ with its standard Euclidean structure, then $\underset{\sim}{K}(R; E^n)_0$ is an n-fold delooping of $\underset{\sim}{K}(R)_0$. The subscript "0" denotes zero-th space. These deloopings are not in general connective; in fact, they are equivalent to the so-called Gersten-Wagner deloopings of $\underset{\sim}{K}(R)_0$. Also, they show that if $X \subseteq S^{n-1}$, and $CX \subseteq E^n$ is the open cone on X, with induced metric, then $\underset{\sim}{K}(R; CX) \cong \Sigma X \wedge \widehat{\underset{\sim}{K}}R$, where $\widehat{\underset{\sim}{K}}R$ is the spectrum whose nth space is $\underset{\sim}{K}(R; E^n)_0$. Finally, it is clear from the construction that if X is a bounded metric space, then $\underset{\sim}{K}(R; X) \cong \widehat{\underset{\sim}{K}}(R)$.

Our generalization of the above construction for $\Gamma = \mathbf{Z}$ now goes as follows. If the metric space X is acted on isometrically by a group Γ, then there is a spectrum $\underset{\sim}{K}_\Gamma(R; X)$, with Γ-action, so that $\underset{\sim}{K}_\Gamma(R; X) \cong \underset{\sim}{K}(R; X)$ non-equivariantly. One can show that if the Γ-action is free, and if X/Γ is bounded, then $\widehat{\underset{\sim}{K}}_\Gamma(R; X)^\Gamma \cong \underset{\sim}{K}(R\Gamma; X/\Gamma) \cong \widehat{\underset{\sim}{K}}(R\Gamma)$. There is an evident induced metric on X/Γ, which is the one we use. We have now achieved our goal of constructing a spectrum with Γ-action whose fixed point spectrum is the K-theory spectrum of $R\Gamma$, if we can find the correct metric space.

Now suppose that $\Gamma = \pi_1 X$, where X is a compact closed manifold. If we equip X with a Riemannian metric, X becomes a bounded metric space. Of course, the Riemannian metric pulls back to the universal cover $\widetilde{X}$, where it is invariant by the action of Γ by deck transformations. Thus $\widetilde{X}$ is a good choice of metric space with free Γ-action. If X were actually flat, then $\widetilde{X} = E^n$, and we would understand $\underset{\sim}{K}(R; E^n)$ from the work of Pedersen and Weibel. In particular, we would find that we have a spectral sequence with $E_2^{p,q} = H^{-p}(\Gamma, K_{q-n}\mathbf{Z})$ converging to $\underset{\sim}{K}_\Gamma(\mathbf{Z}; E^n)^{h\Gamma}$. If we examine the spectral sequence, we find that the groups which appear are precisely those which appear in a similar spectral sequence for $\pi_*(B\Gamma_+ \wedge \underset{\sim}{K}\mathbf{Z})$, because of Poincaré duality in $H^*(\Gamma; \mathbf{Z})$. One can make a precise argument which shows that in fact $\underset{\sim}{K}_\Gamma(\mathbf{Z}; E^n)^{h\Gamma} \cong B\Gamma_+ \wedge \underset{\sim}{K}\mathbf{Z}$, and that the composite $B\Gamma_+ \wedge \underset{\sim}{K}\mathbf{Z} \xrightarrow{\alpha} \underset{\sim}{K}(\mathbf{Z}\Gamma) \to \underset{\sim}{K}_\Gamma(\mathbf{Z}; E^n)^{h\Gamma}$ is an equivalence, allowing us to conclude that α is the inclusion of a wedge factor.

Suppose now $\Gamma = \pi_1 X$, where X is closed, compact, and $\widetilde{X} = \mathbf{R}^n$. In general, the metric will not be flat, and we are unable to use the Pedersen-Weibel results. One can hope, however, to use Mayer-Vietoris techniques to understand $\underset{\sim}{K}(R; \widetilde{X})$ in certain situations. In fact, Pedersen and Weibel used a Mayer-Vietoris sequence in proving their result for Euclidean space. One could ask, then, if a metric space X is decomposed as a union $X = Y \cup Z$, does one have a cofibre sequence

$$\underset{\sim}{K}(R; Y \cap Z) \longrightarrow \underset{\sim}{K}(R; Y) \vee \underset{\sim}{K}(R; Z) \longrightarrow \underset{\sim}{K}(R; X)\,?$$

In general, this fails as one can see from the following picture in the plane E^2.

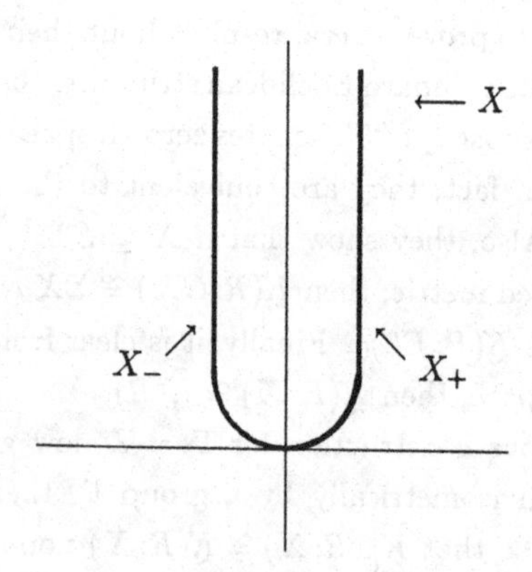

The subset X is equipped with the induced distance function (not the induced Riemannian metric) from E^2. Of course, X is homeomorphic to the real line, and $X = X_+ \cup X_-$. One shows easily that $\underset{\sim}{K}(R; X_+) \cong * \cong \underset{\sim}{K}(R; X_-)$, and of course $\underset{\sim}{K}(R; X_+ \cap X_-) \cong \underset{\sim}{K}(R)$. Thus, if there were a Mayer-Vietoris sequence, we would have a cofibre sequence

$$\underset{\sim}{K}(R; X_+ \cap X_-) \longrightarrow \underset{\sim}{K}(R; X_+) \vee \underset{\sim}{K}(R; X_-) \longrightarrow \underset{\sim}{K}(R; X),$$

and hence that $\underset{\sim}{K}(R; X) \cong \Sigma \underset{\sim}{K}(R; X_+ \cap X_-)$. On the other hand, it is easy to see that projection on the y-axis induces an equivalence $\underset{\sim}{K}(R; X) \to \underset{\sim}{K}(R; \mathbf{R}_+)$, where $\mathbf{R}_+$ denotes the positive y-axis. Pedersen and Weibel have shown this latter space to be contractible, so the sequence cannot be correct as it stands.

Fortunately, it is not too difficult to remedy this situation. If U is a subset of a metric space X, and r is a number, then let $N_r U = \{x \in X \mid \exists u \in U \text{ with } d(x, u) \leq r\}$. Then it turns out that one can construct a Mayer-Vietoris cofibration sequence

$$\varinjlim_r \underset{\sim}{K}(R; N_r U \cap N_r V) \longrightarrow \varinjlim_r \underset{\sim}{K}(R; N_r U) \vee \varinjlim_r \underset{\sim}{K}(R; N_r V) \longrightarrow \underset{\sim}{K}(R; X),$$

when $U \cup V = X$. Using this Mayer-Vietoris sequence, one can now proceed as follows to produce a target for a map out from $\underset{\sim}{K}(R; X)$. Let $\mathcal{U} = \{U_\alpha\}_{\alpha \in A}$ be any finite covering of X. We define a simplicial complex $\beta(\mathcal{U})$ by letting the vertices of $\beta(\mathcal{U})$ be in 1-1 correspondence with A, so $V(\beta(\mathcal{U})) = \{v_\alpha\}_{\alpha \in A}$, and declaring that $\{v_{\alpha_1}, v_{\alpha_2}, \ldots, v_{\alpha_s}\}$ spans a simplex if and only if there is a number r so that the subset $N_r U_{\alpha_1} \cap N_r U_{\alpha_2} \cap \cdots \cap N_r U_{\alpha_s} \subseteq X$ is unbounded. The existence of the above Mayer-Vietoris sequence permits the construction of a map $\underset{\sim}{K}(R; X) \xrightarrow{\tau(\mathcal{U})} \Sigma \beta(\mathcal{U}) \wedge \underset{\sim}{K}(R)$. After some technical work, one obtains a map $\tau: \underset{\sim}{K}(R; X) \to \operatorname{holim}_{\mathcal{U}} \Sigma \beta(\mathcal{U}) \wedge \underset{\sim}{K}(R)$, the homotopy inverse limit over a category of coverings. In fact, this category may be taken to be directed, and the resulting prospectrum is often rather easy to understand.

One can now use slight elaborations of the above ideas to prove that if $\Gamma = \pi_1 X$, where X is closed, compact, and admits a metric of non-positive curvature, then $\alpha: B\Gamma_+ \wedge \underset{\sim}{K}\mathbf{Z} \to \underset{\sim}{K}\mathbf{Z}\Gamma$ is injective onto a summand. The elaboration consists of building the correct Γ-equivariant version of the map τ. The important fact derived from the curvature condition is that two distinct geodesics emanating from the same point grow infinitely far apart. In outline, the proof goes as follows. One constructs a so-called locally-finite homology spectrum of $\widetilde{X}$ with coefficients in $\underset{\sim}{K}\mathbf{Z}$, $h_{\mathrm{lf}}(\widetilde{X}; \underset{\sim}{K}\mathbf{Z})$, which has the following properties.

(a) $h_{\mathrm{lf}}(\widetilde{X}; \underset{\sim}{K}\mathbf{Z})^\Gamma \cong \underset{\sim}{K}\mathbf{Z} \wedge B\Gamma_+$;

(b) $h_{\mathrm{lf}}(\widetilde{X}; \underset{\sim}{K}\mathbf{Z})^{h\Gamma} \cong h_{\mathrm{lf}}(\widetilde{X}; \underset{\sim}{K}\mathbf{Z})^\Gamma$;

(c) $h_{\mathrm{lf}}(\widetilde{X}; \underset{\sim}{K}\mathbf{Z}) \cong S^n \wedge \underset{\sim}{K}\mathbf{Z}$, where $n = \dim X$;

(d) there exists a map $\overline{\alpha}: h_{\mathrm{lf}}(\widetilde{X}; \underset{\sim}{K}\mathbf{Z}) \to \underset{\sim}{K}(\mathbf{Z}; \widetilde{X})$, so that the induced map on fixed points is the assembly map α;

(e) the map $h_{\mathrm{lf}}(\widetilde{X}; \underset{\sim}{K}\mathbf{Z}) \to \underset{\sim}{K}(\mathbf{Z}, \widetilde{X})$ is onto a Γ-equivariantly split summand.

It is in proving part (e) that one uses the map τ above. The curvature condition allows one to choose coverings judiciously, so as to prove (e).

3 Concluding Remarks

(i) One can attack the question of whether α is in fact an equivalence by these methods as well. One must show that the map $\underset{\sim}{K}(\mathbb{Z}\Gamma) \to \underset{\sim}{K}(\mathbb{Z};\widetilde{X})^{h\Gamma}$ is also split injective. This one can do by studying the bounded K-theory of the universal covering space of the stable normal bundle to X. One must also compute precisely $\underset{\sim}{K}(\mathbb{Z},\widetilde{X})$. The precise results one obtains will appear in due course [4].

(ii) The condition that the manifold be closed should be removable in many cases. In particular, I expect that results will be obtained for arithmetic groups.

(iii) The method is not restricted to fundamental groups of manifolds. In particular, the so-called Bruhat-Tits buildings are metric spaces which are not manifolds. The computation of their bounded K-theory should give results on cocompact discrete subgroups over p-adic fields, and S-arithmetic groups.

References

[1] M.F. Atiyah: Characters and cohomology of finite groups. *Inst. Hautes Études Sci. Publ. Math.* 9 (1961), 23–64.

[2] M.F. Atiyah, G.B. Segal: Equivariant K-theory and completion. *J. Differential Geom.* 3 (1969), 1–18.

[3] G. Carlsson: Segal's Burnside ring conjecture and the homotopy limit problem. *Homotopy Theory, edited by E. Rees and J.D.S. Jones, London Mathematical Society Lecture Note Series 117, pp. 6–34, Cambridge University Press*, 1987.

[4] G. Carlsson: In preparation.

[5] T. Farrell, L. Jones: K-theory and dynamics. II. *Ann. of Math.* 126 (1987), 451–493.

[6] E.K. Pedersen, C. Weibel: A non-connective delooping of algebraic K-theory. *Lecture Notes in Math. 1126, pp. 166–181, Springer*, 1985.

[7] D.G. Quillen: Higher algebraic K-theory. I. *Lecture Notes in Math. 341, Springer*, 1973.

[8] F.S. Quinn: Applications of topology with control. *Proceedings of the International Congress of Mathematicians 1986, Berkeley, pp. 598–606.*

[9] R. Thomason: First quadrant spectral sequences in algebraic K-theory via homotopy colimits. *Comm. Algebra* 10 (1982), 1589–1668.

[10] F. Waldhausen: Algebraic K-theory of generalized free products. *Ann. of Math.* 108 (1978), 135–256.

Applications of loop spaces to some problems in topology

F. R. Cohen*

Introduction

This note is an exposition of two related problems. The first problem concerns the cohomology of mapping class groups and a relation to function spaces. The second problem concerns the existence of a bound on the order of the torsion in the homotopy groups of 1-connected mod-2^r Moore spaces. These two problems have common themes: the double loop space of a sphere and the Whitehead square. We would like to thank the organizers of this conference, especially Wilson Sutherland, for providing such a stimulating and delightful environment.

1 Mapping class groups and function spaces

In this section we explore a relation between certain mapping class groups, function spaces, and their cohomology. Much of the work here is an outgrowth of combinatorial models first given by James [23], Dyer and Lashof [18], and others [1,7,25,26,29]. In particular, these models provide a systematic procedure for determining the cohomology of certain groups from the cohomology of function spaces which in turn can be computed by other continuous methods.

Let $M^n_{g,k}$ denote an orientable Riemann surface of genus g with k boundary components and n punctures. The mapping class group $\Gamma^n_{g,k}$ is $\pi_0 \mathrm{Diff}^+(M^n_{g,k})$ where $\mathrm{Diff}^+(\cdot)$ is the group of orientation preserving diffeomorphisms fixing the boundary circles pointwise. Write $M_g = M^0_{g,0}$, $\Gamma_g = \Gamma^0_{g,0}$ and $\Gamma^n = \Gamma^n_{0,0}$. Work of Earle and Eells [19] gives that each component of $\mathrm{Diff}^+(M_g)$ is contractible and thus $B\Gamma_g$ classifies bundles with fibre M_g for $g \geq 2$. Some beautiful results on the cohomology of $\Gamma^n_{g,k}$ are given in work of Harer [22] and E. Miller [27]. It is pointed out in [10] that the classifying space $B\Gamma$ where $\Gamma = \varinjlim \Gamma_{g,1}$ has the space $\mathrm{Im} J_{(\frac{1}{2})}$ as a stable retract. Other related results are given in [2,3,14].

*Partially supported by the National Science Foundation

Classically, the surface M_g can be regarded as a 2-sheeted branched cover of S^2, $\pi: M_g \to S^2$, as follows [5]. There is an involution $j: M_g \to M_g$ with $2g+2$ fixed points and S^2 is the orbit space of the induced $\mathbb{Z}/2\mathbb{Z}$-action on M_g. Define the hyperelliptic mapping class group Δ_g to be the centralizer of the class of j in Γ_g.

By work of Bergau and Mennicke [4] and Birman and Hilden [6], we have

(1) $\Delta_2 = \Gamma_2$,

(2) Δ_g is a proper subgroup of Γ_g if $g \geq 3$, and

(3) there are non-split central extensions $1 \to \mathbb{Z}/2\mathbb{Z} \to \Delta_g \to \Gamma^{2g+2} \to 1$.

In this section we describe some joint work of C.-F. Bödigheimer, M.D. Peim, and the author concerning the groups Γ^n and consequently Δ_g [8].

There are very simple constructions for the spaces $K(\Gamma^n, 1)$. Start with a fibre bundle $\pi: E \to B$ with fibre Y. Fadell and Neuwirth defined configuration spaces in [20] and their bundle analogues were studied in [17]. Namely write

$$F(Y,k) = \{(y_1, \ldots, y_k) \in Y^k \mid y_i \neq y_j \text{ if } i = j\}$$

and

$$E(\pi,k) = \{(e_1, \ldots, e_k) \in E^k \mid e_i \neq e_j \text{ and } \pi(e_i) = \pi(e_j) \text{ if } i = j\}.$$

Thus there are bundles

$$F(Y,k) \longrightarrow E(\pi,k) \longrightarrow B$$

and

$$F(Y,k)/\Sigma_k \longrightarrow E(\pi,k)/\Sigma_k \longrightarrow B,$$

where the symmetric group Σ_k acts by permuting coordinates.

Consider the bundle specified by $\eta: BSO(2) \to BSO(3)$ with fibre S^2. Of course we take $BSO(2)$ to be $ESO(3)/SO(2)$ where $SO(2)$ is naturally embedded in $SO(3)$. Much of what we say below generalizes, but this example will suffice here.

1.1 **Lemma** [8]. *If $k \geq 3$, then $E(\eta,k)/\Sigma_k$ is a $K(\Gamma^k, 1)$ and so there is a fibration*

$$F(S^2,k)/\Sigma_k \longrightarrow K(\Gamma^k,1) \longrightarrow BSO(3).$$

Sketch. There is a fibration

$$F(\mathbb{R}^2, k-1) \longrightarrow E(\eta,k) \longrightarrow BSO(2)$$

and so $\pi_i E(\eta,k) = 0$ if $i \geq 2$ by inspecting the boundary $\pi_2 BSO(2) \to \pi_1 F(\mathbb{R}^2, k-1)$. To compute the fundamental group of $E(\eta,k)/\Sigma_k$ inspect the long exact homotopy sequence for the fibration $F(S^2,k)/\Sigma_k \to E(\eta,k)/\Sigma_k \to BSO(3)$ together with [21]. ∎

The space $E(\eta,k)/\Sigma_k$ admits an equivalent interpretation analogous to a Grassmannian [3]. Fix a Euclidean space $\mathbf{R}^L$ and write $A(\mathbf{R}^L,k)$ to be the space of ordered k-tuples $((v_1,P_1),\ldots,(v_k,P_k))$ where

(1) $v_1,\ldots,v_k$ are distinct unit vectors in $\mathbf{R}^L$,

(2) P_i is a 2-plane orthonormal to v_i, and

(3) the 3-planes spanned by v_i and P_i coincide for all i.

There are inclusions of $A(\mathbf{R}^L,k)$ in $A(\mathbf{R}^{L+1},k)$ and the union of $A(\mathbf{R}^L,k)/\Sigma_k$ is $K(\Gamma^k,1)$.

1.2 **Lemma** [3]. *If $n \geq 6$, the map $K(\Gamma^n,1) \longrightarrow BSO(3)$ in 1.1 is not homotopic to $B(g)$ where g is a homomorphism from Γ^n to $SO(3)$.*

Sketch. There is a presentation of Γ^n where all generators are conjugate [24]. That many generators commute if $n \geq 7$ immediately gives too many non-trivial distinct commuting elements in $SO(3)$ as such elements are given by rotations in orthogonal hyperplanes. If $n = 6$, then g factors through a cyclic group and this contradicts Theorem 1.6 here. ■

The symmetric groups Σ_n are natural quotients of Γ^n while Γ^n is a natural quotient of Artin's braid group B_n [24]. It thus seems natural to construct stabilization maps from Γ^n to Γ^{n+1}. It is easy to see that one cannot do this on the level of groups so that the homomorphisms are compatible with the stabilization maps for the symmetric groups. Indeed the maps cannot even exist on the level of cohomology by inspecting Theorem 1.6 here. In what follows we overcome this technical difficulty by the function space method suggested in the introduction.

Here it is convenient to mimic combinatorial models as follows: Given a pointed space X with basepoint $*$, define $E(\pi;X)$ to be the equivalence classes of pairs $[S,f]$ where

(0) $\pi\colon E \to B$ is a bundle projection,

(1) S is a finite subset of E and $\pi(S)$ has cardinality 1,

(2) $f\colon S \to X$, and

(3) $[S,f]$ is equivalent to $[S-\{q\}, f|S-\{q\}]$ if and only if $f(q) = *$.

Thus there is a bundle $C(Y;X) \to E(\pi;X) \to B$ where $\pi\colon E \to B$ is a bundle projection with fibre Y and $C(Y;X)$ is $E(\lambda;X)$ where $\lambda\colon Y \to *$ is the constant map. The space $E(\pi;X)$ is filtered by the cardinality of S and $D_k(\pi;X)$ denotes the associated filtration quotients. Assume throughout that X is a CW-complex.

In the special case of $\eta\colon BSO(2) \to BSO(3)$, $E(\eta;X)$ is just the Borel construction $ESO(3) \times_{SO(3)} C(S^2;X)$ where $SO(3)$ acts by rotations on S^2. If $X = \Sigma Y$, then $C(S^2;X)$ is homotopy equivalent to the space $\Lambda^2\Sigma^2 X$ of all (unpointed) maps from S^2 to $\Sigma^2 X$. This is essentially proven in [7] with a technical modification in [8]. In addition, the natural map $C(S^2;X) \to \Lambda^2\Sigma^2 X$ is $SO(3)$-equivariant.

1.3 Proposition [8]. *If $X = \Sigma Y$ is a CW-complex, then there is a map of fibrations*

$$\begin{array}{ccc} E(\eta, X) & \xrightarrow{\phi} & ESO(3) \times_{SO(3)} \Lambda^2\Sigma^2 X \\ \downarrow & & \downarrow \\ BSO(3) & \xrightarrow{1} & BSO(3), \end{array}$$

where ϕ is a homotopy equivalence.

The following result follows immediately from the proof in the appendix of [12] although the result is not stated explicitly there.

1.4 Proposition. *If X is a connected CW-complex, then $E(\pi; X)$ is stably homotopy equivalent to $\bigvee_{k\geq 1} D_k(\pi; X)$.*

1.5 Remark. Proposition 1.4 provides a method to determine $H_*(\Gamma^n; V^{\otimes n})$ where V is a graded vector space over F concentrated in degrees greater than 0 and Γ^n acts on $V^{\otimes n}$ by permuting coordinates with the usual sign conventions. If S_V is a bouquet of spheres with $\bar{H}_*(S_V; \mathsf{F})$ isomorphic to V, then there is an isomorphism

$$H_*(\Gamma^n; V^{\otimes n}) \cong H_*(D_n(\eta; S_V); \mathsf{F}).$$

Thus if F denotes the trivial Γ^n-module and $\mathsf{F}(-1)$ denotes the sign representation, there are isomorphisms

$$H_q(\Gamma^n; \mathsf{F}) \cong H_{q+2jn} D_n(\eta; S^{2j}),$$

and

$$H_q(\Gamma^n; \mathsf{F}(-1)) \cong H_{q+(2j+1)n} D_n(\eta; S^{2j+1}).$$

Thus it suffices to know the homology of $E(\eta; S_V)$ as a filtered vector space to determine the additive structure of $H_*(\Gamma^n; V^{\otimes n})$. We shall carry this out in the special case where $V = \mathsf{F}_2$. Other cases are worked out in [8].

Consider the bundle

$$S^2 \times S^n \longrightarrow BSO(2) \times S^n \longrightarrow BSO(3)$$

obtained from η to get a map of bundles

$$\begin{array}{ccc} BSO(2) \times S^n & \longrightarrow & E(\eta; S^n) \\ \downarrow & & \downarrow \\ BSO(3) & \xrightarrow{1} & BSO(3). \end{array}$$

By naturality, there is a non-trivial differential in the cohomology Serre spectral sequence for the right-hand fibration. There is a natural map of $E(\eta; S^n)$ to $\Omega^\infty S^{\infty+n}$ and this directly implies that the d_3-differential given above is the only non-trivial differential. As a consequence we deduce the following result.

1.6 Theorem [8]. *There are isomorphisms of $H^*(BSO(3);F_2)$-modules*

(1) $H^*(\Gamma^{2n};F_2) \to H^*(BSO(3);F_2) \otimes H^*(F(S^2,2n)/\Sigma_{2n};F_2)$ for $n \geq 2$, and

(2) $H^*(\Gamma^{2n+1};F_2) \to H^*(BSO(2);F_2) \otimes H^*(F(R^2,2n+1)/\Sigma_{2n+1};F_2)$ for $n \geq 1$.

We remark that $H^*(F(S^2,n)/\Sigma_n;F_2)$ is given in [9]. There are vector space isomorphisms

(1) $H^*(F(S^2,2n)/\Sigma_{2n};F_2) \cong H^*(B_{2n};F_2) \oplus H^{*-2}(B_{2n-2};F_2)$ and

(2) $H^*(F(S^2,2n+1)/\Sigma_{2n+1};F_2) \cong H^*(B_{2n+1};F_2) \oplus H^{*-2}(B_{2n-1};F_2)$ where B_i is Artin's braid group on i strings.

The calculations at odd primes depend on the analysis of the composite

$$\Omega^2 S^{4n+3} \xrightarrow{\Omega^2 w} \Omega^2 S^{2n+2} \longrightarrow \Lambda^2 S^{2n+2} \longrightarrow ESO(3) \times_{SO(3)} \Lambda^2 S^{2n+2}$$

in homology where w is the Whitehead square. We refer the reader to [8] for details. If n is small, some previous calculations for $H^*(\Gamma^n, F_p)$ were given in [14]. The mod-2 and mod-3 cohomology of Γ^6 was first determined in work of Dave Benson [2] who used the methods in [14] where the cohomology of Γ^n was determined for various values of n. The methods are different and involve the analysis of a Lyndon-Hochschild-Serre spectral sequence with twisted coefficients.

2 On the homotopy theory of mod-2^r Moore spaces

Throughout this section $P^n(p^r)$ denotes the Moore space given by the cofibre of a degree p^r map $[p^r]: S^{n-1} \to S^{n-1}$. The symbol p^r will be reserved for the p^r-th power map on a loop space ΩX. In the papers [15,16,28], the homotopy theory of odd primary Moore spaces was studied by constructing various product decompositions. The paucity of elements of either Hopf invariant one or Kervaire invariant one is an obstruction to the existence of analogous splittings when $p = 2$. Here we amalgamate classical techniques arising from James-Hopf invariants together with splitting techniques to prove the following result.

2.1 Theorem. *If $n \geq 3$ and $r \geq 2$, then 2^{2r+4} annihilates $\pi_* P^n(2^r)$.*

If p is an odd prime and $n \geq 3$, then p^{r+1} annihilates $\pi_* P^n(p^r)$ [28]. Some previous information was given in [16]. In [13] certain elements of order 2^{r+1} in $\pi_* P^n(2^r)$ were constructed. It follows immediately that there are infinitely many elements of order 2^{r+1} in $\pi_* P^n(2^r)$ if $r \geq 2$ and $n \geq 3$. Unfortunately these elements of order 2^{r+1} are not analogous to the "canonical" elements of order p^{r+1} in $\pi_* P^n(p^r)$ given in [15] for p odd.

It seems quite likely that the techniques here will give that 2^{r+3} annihilates the homotopy groups of $P^n(2^r)$ if $r \geq 2$ and $n \geq 3$. However, this is still work in progress. M.G. Barratt conjectures that 2^{r+1} should annihilate $\pi_* P^n(2^r)$ if $r \geq 2$ and $n \geq 3$.

The first step is to construct non-trivial maps from loop spaces to $S^{2n+1}\{2^r\}$, the fibre of the map $[2^r]: S^{2n+1} \to S^{2n+1}$. If $S^{2n+1}\{2^r\}$ were an H-space, such constructions would be easy.

2.2 Lemma. *If $n \geq 1$, then $S^{2n+1}\{2\}$ is not an H-space. If the Whitehead square $[\imath_{2n+1}, \imath_{2n+1}]$ in $\pi_{4n+1}S^{2n+1}$ is not divisible by 2, then $S^{2n+1}\{2^r\}$ is not an H-space.*

Sketch. If $S^{2n+1}\{2^r\}$ were an H-space, then there is a degree one map $\Omega P^{2n+2}(2^r) \to S^{2n+1}\{2^r\}$. Thus $\Omega P^{2n+2}(2^r)$ would split as $S^{2n+1}\{2^r\}\times$ another space. If $r = 1$, and $n \geq 1$, this contradicts the action of Sq^2. In general, this splitting forces the square v^2 in $H_{4n+2}(\Omega P^{2n+2}(2^r); \mathbb{F}_2)$ to be mod-2^r spherical. Thus a similar homological assertion holds for $H_{4n+2}(\Omega S^{2n+2}; \mathbb{F}_2)$ and so $[\imath_{2n+1}, \imath_{2n+1}]$ is divisible by 2 [30]. We remark that $S^{2n+1}\{2^r\}$ is sometimes an H-space. ∎

Let $f: X \to Y$ be a *fixed* map. We say that the triple (X, Y, f) is a partial H-space provided there is a homotopy commutative diagram

$$\begin{array}{ccc} X & \xrightarrow{f} & Y \\ E\downarrow & & \uparrow\theta \\ \Omega\Sigma X & \xrightarrow{1} & \Omega\Sigma X. \end{array}$$

Of course partial H-spaces occur ubiquitously as can be seen by considering fibrations with fibre Y and base ΣX. There is a different type of example which is useful in Theorem 2.1.

Consider the natural inclusion $i: P^{2n+1}(2) \to S^{2n+1}\{2\}$. By Lemma 2.2, the pair $(P^{2n+1}(2), S^{2n+1}\{2\}, i)$ is never a partial H-space if $n \geq 1$. Let $\rho: P^{2n+1}(2^r) \to P^{2n+1}(2)$ be induced by mod-2 reduction.

2.3 Lemma. *If $r \geq 2$, then $(P^{2n+1}(2^r), S^{2n+1}\{2\}, \rho\cdot i)$ is a partial H-space.*

Sketch. Let $\tau(S^{2n+2})$ be the unit tangent bundle of S^{2n+2} and let $h_2: \Omega S^{2n+2} \to \Omega S^{4n+3}$ be the second James-Hopf invariant. The fibre of the composite $g: \Omega\tau(S^{2n+2}) \to \Omega S^{4n+3}$ is $S^{2n+1}\{2\}$ where $g = h_2 \cdot \Omega(\pi)$ and $\pi: \tau(S^{2n+2}) \to S^{2n+2}$ is the natural projection map. If $r \geq 2$, then $h_2 \cdot \Omega(\rho \cdot j)$ is null where $j: P^{2n+2}(2) \to \tau(S^{2n+2})$ is the natural inclusion and $\rho: P^{2n+2}(2^r) \to P^{2n+2}(2)$ is given by mod-2 reduction. ∎

2.4 Remark. A modification of the above proof gives a homotopy commutative diagram

$$\begin{array}{ccc} \Omega S^{2n+1} & \xrightarrow{1} & \Omega S^{2n+1} \\ \downarrow & & \downarrow \\ \Omega P^{2n+2}(2^r) & \xrightarrow{\theta} & S^{2n+1}\{2\}, \end{array}$$

where the vertical arrows are obtained by (1) looping the natural map $S^{2n+1} \to P^{2n+2}(2^r)$ and by (2) the action of ΩS^{2n+1} on $S^{2n+1}\{2\}$ gotten from the fibration $[2]\colon S^{2n+1} \to S^{2n+1}$. In addition, there is a homotopy commutative diagram

$$\begin{array}{ccc} \Omega^2 S^{4n+3} & \longrightarrow & S^{2n+1}\{2\} \\ 1\downarrow & & \downarrow \\ \Omega^2 S^{4n+3} & \xrightarrow{\Delta} & S^{2n+1}, \end{array}$$

where Δ induces the boundary map in the classical EHP sequence.

We require some information concerning James-Hopf invariants which are special for double suspensions. Let $h_2\colon \Omega\Sigma X \to \Omega\Sigma X^{(2)}$ be the second James-Hopf invariant where $X^{(j)}$ denotes the j-fold smash product. Let $w_k\colon \Sigma X^{(k)} \to \Sigma X$ be the $(k-1)$-fold Whitehead product $\underbrace{[1[1\cdots[1,1]}_{k}\cdots]$.

2.5 Lemma. *If $X = \Sigma Y$ and Y is a connected CW-complex, then the composite $\Omega\Sigma X^{(k)} \xrightarrow{\Omega^2 w_k} \Omega\Sigma X \xrightarrow{h_2} \Omega\Sigma X^{(2)}$ is homotopic to a loop map.*

Sketch. Since X is a suspension, the inverse map on the James construction JX can be given by sending $(x_1,\ldots,x_n)$ to $(\sigma x_n,\ldots,\sigma x_1)$ where σ is the inverse in the suspension coordinate. The lemma then follows by inspecting the composite $JX^{(k)} \xrightarrow{\Omega w_k} JX \xrightarrow{h_2} JX^{(2)}$ where h_2 is the standard combinatorially defined James-Hopf invariant. We remark that this lemma is false if X is not a suspension. ■

Throughout the rest of this section all homology groups are taken with coefficients in F_2.

2.6 Lemma. *If $r \geq 2$ and $n \geq 3$, there exist spaces $T^n\{2^r\}$ such that*

(1) there is a homotopy equivalence $\Omega P^n(2^r) \to T^n\{2^r\} \times \Omega\Big(\bigvee_{m\in I} P^m(2^r)\Big)$, $m > n$,

and

(2) there is an isomorphism of coalgebras $H_*T^n\{2^r\} \to H_*\Omega S^n \otimes H_*\Omega^2 S^n$.

Sketch. Since $r \geq 2$, the space $P^a(2^r) \wedge P^b(2^r)$ is homotopy equivalent to $P^{a+b}(2^r) \vee P^{a+b-1}(2^r)$. By inspecting [15], we see that there is a multiplicative map $\Omega\phi\colon \Omega\Big(\bigvee_{m\in I} P^m(2^r)\Big) \to \Omega P^n(2^r)$ where $(\Omega\phi)_*$ is a monomorphism with algebra cokernel isomorphic to $H_*\Omega S^n \otimes H_*\Omega^2 S^n$. This suffices by [16]. ■

Recall that if $r \geq 2$ and $n \geq 3$, then $H_*\Omega P^n(2^r)$ is isomorphic to the tensor algebra $T[u,v]$ as Hopf algebra where v has degree $n-1$ and the rth Bockstein of v is u. The homology of $T^n\{2^r\}$ is isomorphic to the polynomial algebra $\mathsf{F}_2[v, \tau_0, \tau_1, \cdots]$ as a coalgebra where $\tau_0 = u$ and $\tau_i = ad^{2^i-1}(v)(u)$.

2.7 Proposition. *If $r \geq 2$ and $n \geq 3$, then there are fibrations*

$$S^{n-1}\{2^r\} \longrightarrow T^n\{2^r\} \longrightarrow T^{2n-1}\{2^r\}.$$

Furthermore if the Whitehead square $[\imath_{n-1}, \imath_{n-1}]$ in $\pi_{2n-1}S^n$ is not divisible by 2, then $T^n\{2^r\}$ is atomic.

Sketch. Since the space $\Sigma T^n\{2^r\}$ is homotopy equivalent to a wedge of $P^{2n-1}(2^r)$ and other Moore spaces, we obtain $h: T^n\{2^r\} \to T^{2n-1}\{2^r\}$. Furthermore $H^*T^n\{2^r\}$ is a free $H^*T^{2n-1}\{2^r\}$-module and so there is a map of $S^{n-1}\{2^r\}$ to the fibre of h which gives an isomorphism in cohomology. The proof of the second statement is analogous to that given in [16]. ■

Finally, we fibre the spaces $T^n\{2^r\}$ into identifiable pieces. Here consider the second James-Hopf invariant $h_2: \Omega P^n(2^r) \to \Omega\Sigma(P^{n-1}(2^r) \wedge P^{n-1}(2^r))$. Pinching $P^{n-1}(2^r) \wedge P^{n-1}(2^r)$ to $P^{2n-2}(2^r)$, we get a map $h: \Omega P^n(2^r) \to \Omega P^{2n-1}(2^r)$ and the iterates $h^j: \Omega P^n(2^r) \to \Omega P^{2^j(n-1)+1}(2^r)$. If $r \geq 2$, we use a choice of degree one map $\pi: P^{n-1}(2^r) \wedge P^{n-1}(2^r) \to P^{2n-3}(2^r)$ and Lemma 2.3 to obtain maps ϕ_{j+1} as follows.

$$\begin{array}{ccccc}
\Omega P^n(2^r) & \xrightarrow{h^j} & \Omega P^{2^j(n-1)+1}(2^r) & \xrightarrow{h_2} & \Omega\Sigma P^{2^j(n-1)}(2^r)^{(2)} \\
 & & & & \downarrow \Omega(\pi) \\
\downarrow 1 & & & & \Omega\Sigma P^{2^{j+1}(n-1)-1}(2^r) \\
 & & & & \downarrow \theta \\
\Omega P^n(2^r) & & \xrightarrow{\phi_{j+1}} & & S^{2^{j+1}(n-1)-1}\{2\}.
\end{array}$$

Thus there are induced maps

$$\Phi: T^{2n+1}\{2^r\} \longrightarrow \prod_{k\geq 2} S^{2^k n-1}\{2\},$$

where $\prod$ is the weak infinite product.

Let $V^{2n+1}\{2^r\}$ denote the homotopy theoretic fibre of the following composite q, $T^{2n+1}\{2^r\} \to \Omega P^{2n+1}(2^r) \to \Omega S^{2n+1}$.

2.8 Theorem. *If $n \geq 1$ and $r \geq 2$, there is a fibration*

$$\Omega S^{2n-1} \times \Omega(\prod_{k\geq 2} S^{2^k n-1}\{2^r\}) \longrightarrow \Omega V^{2n+1}\{2^r\} \longrightarrow \Omega(\prod_{k\geq 2} S^{2^k n-1}\{2\}).$$

Sketch. The image of Φ in homology is isomorphic to $\mathbb{F}_2[\tau_0^2, \tau_1^2, \cdots]$ as a coalgebra, thus the homology of the fibre of q is isomorphic to that of $S^{2n-1} \times \prod_{k\geq 2} S^{2^k n-1}\{2^r\}$ by using the Bockstein lemma. There are clearly maps of $S^{2^k n-1}\{2^r\}$ to $V^{2n+1}\{2^r\}$ exhibiting the correct homological behavior. By Lemmas 2.3, 2.5 and Remark 2.4 these maps factor through the fibre of q. Since q is not an H-map, we cannot directly extend these maps over products. After looping, we obtain a multiplicative fibration and thus a map of $\Omega(S^{2n-1} \times \prod_{k\geq 2} S^{2^k n-1}\{2^r\})$ to the homotopy theoretic fibre of $\Omega(q)$; this last map is a homology isomorphism. ∎

Analogous fiberings hold for $T^{2n+2}\{2^r\}$, $n \geq 1$. An inspection of the long exact homotopy sequence for a fibration gives the following result.

2.9 Corollary. *Assume that $n \geq 1$ and $r \geq 2$. Then*

1. $16\pi_* S^{2n+1}\{2\} = 0$,
2. $2^{2r+4}\pi_* T^{2n+1}\{2^r\} = 0$, *and*
3. $2^{2r+4}\pi_* T^{2n+2}\{2^r\} = 0$.

Notice that Lemma 2.6 and Corollary 2.9 imply Theorem 2.1.

References

[1] M.G. Barratt and P.J. Eccles: Γ^+-structures. III. *Topology* 13 (1974), 199–207.

[2] D. Benson: Specht modules and the cohomology of mapping class groups. To appear.

[3] D. Benson and F.R. Cohen. To appear.

[4] P. Bergau and J. Mennicke: Über topologische Abbildungen der Brezelfläche vom Geschlecht 2. *Math. Z.* 74 (1960), 414–435.

[5] J. Birman: Braids, Links, and Mapping Class Groups. Ann. of Math. Stud. 82, Princeton University Press, 1974.

[6] J. Birman and H. Hilden: Isotopies of homeomorphisms of Riemann surfaces. *Ann. of Math.* 97 (1973), 424–439.

[7] C.-F. Bödigheimer: Stable Splitting of Mapping Spaces. *Lecture Notes in Math. 1286, pp. 174–187, Springer*, 1987.

[8] C.-F. Bödigheimer, F.R. Cohen, and M.D. Peim: Mapping class groups and function spaces. To appear.

[9] C.-F. Bödigheimer, F.R. Cohen, and L.R. Taylor: On the homology of configuration spaces. *Topology* 28 (1989), 111–123.

[10] R. Charney and F.R. Cohen: A stable splitting for the mapping class group. *Michigan Math. J.* 35 (1988), 269–284.

[11] F.R. Cohen: Homology of C_{n+1}-spaces: *Lecture Notes in Math. 533, pp. 207–351, Springer*, 1976.

[12] F.R. Cohen: The unstable decomposition of $\Omega^2\Sigma^2 X$ and its applications. *Math. Z.* 182 (1983), 553–568.

[13] F.R. Cohen: A course in some aspects of classical homotopy theory. *Lecture Notes in Math. 1286, pp. 1–92, Springer*, 1987.

[14] F.R. Cohen: Artin's braid group and the homology of certain subgroups of the mapping class group. To appear.

[15] F.R. Cohen, J.C. Moore and J.A. Neisendorfer: Torsion in homotopy groups. *Ann. of Math.* 109 (1979), 121–168.

[16] F.R. Cohen, J.C. Moore, and J.A. Neisendorfer: Exponents in homotopy theory. *Ann. of Math. Stud.* 113 (1987), 3–34.

[17] F.R. Cohen and L.R. Taylor: Computations of Gelfand-Fuks cohomology, the cohomology of function spaces, and the cohomology of configuration spaces. *Lecture Notes in Math. 657, pp. 106–143, Springer*, 1978.

[18] E. Dyer and R. Lashof, Homology of iterated loop spaces: *Amer. J. Math.* 84 (1962), 35–88.

[19] C. Earle and J. Eells: A fibre bundle description of Teichmüller theory. *J. Differential Geom.* 3 (1969), 19–43.

[20] E. Fadell and L. Neuwirth: Configuration spaces. *Math. Scand.* 10 (1963), 111–118.

[21] E. Fadell and J. Van Buskirk: The braid groups of E^2 and S^2. *Duke Math. J.* 29 (1963), 243–258.

[22] J. Harer: Stability of the homology of the mapping class groups of orientable surfaces. *Ann. of Math.* 121 (1985), 215–249.

[23] I.M. James: Reduced product spaces. *Ann. of Math.* 62 (1955), 170–197.

[24] W. Magnus: Über Automorphismen von Fundmentalgruppen berandeter Fläche. *Math. Ann.* 109 (1934), 617–646.

[25] J.P. May: The Geometry of Iterated Loop Spaces. Lecture Notes in Math. 271, Springer, 1972.

[26] R.J. Milgram: Iterated loop spaces. *Ann. of Math.* 84 (1966), 386–403.

[27] E. Miller: The homology of the mapping class group. *J. Differential Geom.* 24 (1986), 1–14.

[28] J.A. Neisendorfer: The exponent of a Moore space. *Ann. of Math. Stud.* 113 (1987), 35–71.

[29] G. Segal: Configuration spaces and iterated loop spaces. *Invent. Math.* 21 (1973), 213–221.

[30] P.S. Selick: A reformulation of the Arf invariant one mod p problem and applications to atomic spaces. *Pacific J. Math.* 108 (1983), 431–450.

Invariants of fixed-point-free circle actions

M. C. Crabb

Introduction

Stable homotopy invariants of fixed-point-free circle actions appear as natural tools in various geometric problems. Here I treat one problem as an example. (Others, some joint work with A.J.B. Potter and B. Steer, include applications to the indices of vector fields and flows and to the $\bar{\eta}$-invariant of the Dirac operator.)

The content of Section 1 is mostly fairly standard equivariant homotopy theory specialized to the circle group, $\mathbf{T}$. Section 2 contains more detailed calculations. (I have tried to make these sections accessible, at least at a superficial level, to the reader who is not an expert in equivariant homotopy theory and, where possible, have included a few lines of explanation, in the context of circle actions, rather than simply a reference to a general theorem. Some of this material will be needed for the other applications mentioned above.)

The problem treated in the last three sections is the following. Let M be a framed (closed) manifold. It has been well known since the work of Knapp [19] that, if M admits a free $\mathbf{T}$-action compatible with the framing, then the complex e-invariant of M is determined by an equivariant K-theory invariant one degree simpler than the e-invariant. This result is generalised in Section 4, 4.3, from free to fixed-point-free $\mathbf{T}$-actions. In Section 5 it is shown that in certain circumstances the K-theory invariant can be computed from the Euler characteristics $\chi(M^{\mathbf{T}(k)}/\mathbf{T})$ (where $M^{\mathbf{T}(k)}$ is the submanifold fixed by the kth roots of unity in $\mathbf{T}$). The result [22] of Seade and Steer on the e-invariant of the quotient $\mathbf{PSL}(2,\mathbf{R})/\Gamma$ by a co-compact Fuchsian group Γ is derived as an illustration. (It was this example which motivated the generalities of Section 4.) Section 3 contains preliminary observations on the geometry.

Notation. The circle group, here thought of as the group of complex numbers of modulus 1, is written as $\mathbf{T}$, and the subgroup of kth roots of unity, $k \geq 1$, as $\mathbf{T}(k)$. The basic 1-dimensional representation of $\mathbf{T}$, $\mathbf{C}$ with $\mathbf{T}$ acting by multiplication, is denoted by E; tensor powers are written as E^k, $k \in \mathbf{Z}$.

An action of a group G on a space X is called *fixed-point-free* if the fixed subspace X^G is empty, or, in other words, if every isotropy group is a proper subgroup of G. For the circle, $G = \mathbf{T}$, this is the same as an action with finite isotropy groups. (The action is *free* if every isotropy group is trivial.)

Acknowledgments. My interest in $\mathbf{T}$-equivariant homotopy group theory dates from the beginning of joint work with K. Knapp on the stable homotopy of complex projective spaces almost a decade ago. I would like to record my thanks to him for many conversations related, at least indirectly, to the subject of this paper. My thanks are also due to I. M. James for his encouragement and numerous invitations to Oxford, where I have had many fruitful discussions.

1 Preparation

This section establishes notation and reviews some basic concepts of $\mathbf{T}$-equivariant homotopy theory. It begins with a catalogue of the various cohomology theories that we shall need.

The first is *stable homotopy*, which will be denoted by the somewhat unorthodox "ω"(1). Let A, B be compact ENR's with basepoint. We write $\omega^0\{A; B\}$ for the abelian group of stable maps $A \to B$ and more generally, using cohomology indexing, for any integer $i \in \mathbf{Z}$:

$$\omega^i\{A; B\} = \omega^0\{\Sigma^{N-i}A; \Sigma^N B\} \quad \text{where } N \geq \max\{0, i\}. \tag{1.1}$$

In particular,

$$\omega^i\{A; S^0\} = \tilde{\omega}^i(A) \tag{1.2}$$

is the *reduced* stable cohomotopy of A, and

$$\omega^{-j}\{S^0; B\} = \tilde{\omega}_j(B) \tag{1.3}$$

the reduced stable homotopy of B. If X is a compact ENR, then X_+ will denote the pointed space obtained by adjoining a disjoint basepoint to X. We write

$$\omega^i(X) = \tilde{\omega}^i(X_+) \quad \text{and} \quad \omega_j(X) = \tilde{\omega}_j(X_+) \tag{1.4}$$

for the unreduced groups(2). In its simplest form stable homotopy theory is the study of the category in which the objects are (finite complexes or) compact ENR's with

(1) In this I am following equivariant homotopy theorists, notably tom Dieck [14].

(2) The notation $\omega^i\{A; B\}$ is compatible with the widely used $\{A; B\}$ for stable maps. The form $\omega^i(A; B)$, sometimes read as the stable cohomotopy of A with coefficients in B, is more common (and preferred in [14]). In geometric problems where unreduced cohomology groups arise naturally it might be more logical to affix a tilde; tom Dieck does this in [12].

basepoint and $\omega^0\{A,B\}$ is the set of morphisms from A to B. To define manageable invariants one usually has to pass to a more readily computable multiplicative cohomology theory and its associated category. In this paper we shall work with (periodic) K-theory and classical cohomology. Notation will correspond exactly to that already introduced in (1.1)–(1.4) for stable cohomotopy.

Periodic complex K-theory will be denoted by "K". The group $K^0\{A;B\}$ is most succinctly defined using maps of spectra as $[\mathbf{A};\mathbf{B}\wedge\mathbf{K}]$, where $\mathbf{A}$ and $\mathbf{B}$ are the suspension spectra of A and B and $\mathbf{K}$ is the representing spectrum of periodic complex K-theory. Kasparov's KK definition (described, for example, in [7]) is probably more illuminating, as is Segal's definition of connective k-theory, [23]. In any case, one should think of $K^0\{A;B\}$ as the *K-theory maps from A to B*, the morphisms $A\to B$ in the K-theory category. Then the Hurewicz construction defines a natural transformation from the stable homotopy to the K-theory category, and this transformation is multiplicative: that is, compatible with smash products.

The letter "H" without qualification will denote *classical cohomology with integer coefficients*. The group $H^0\{A;B\}$ can be defined formally using the Eilenberg-MacLane spectrum. Again, other definitions are more interesting. For example, in singular homology theory it would be natural to introduce $H^0\{A;B\}$ as the group of chain maps $\tilde{S}_*(A)\to\tilde{S}_*(B)$ from the (reduced) singular chain complex of A to that of B, [11]. Rationally we have simply:

$$H^0\{A;B\}\otimes\mathbf{Q}=\prod_{i\in\mathbf{Z}}\operatorname{Hom}\left(\tilde{H}_i(A;\mathbf{Q}),\tilde{H}_i(B;\mathbf{Q})\right),\tag{1.5}$$

and this determines $\omega^0\{A;B\}\otimes\mathbf{Q}$.

To discuss the *Chern character* it is convenient to define a cohomology theory "$\mathfrak{C}$" by:

$$\mathfrak{C}^0\{A;B\}=\bigoplus_{j\in\mathbf{Z}}H^{2j}\{A;B\}\otimes\mathbf{Q}\tag{1.6}$$

with the usual product (as in 1.21 below). In particular, we have the familiar cohomology group

$$\tilde{\mathfrak{C}}^i(A)=\bigoplus_{j\in\mathbf{Z}}\tilde{H}^{i+2j}(A;\mathbf{Q}).$$

The four theories are related by the following commutative square of multiplicative natural transformations.

$$\begin{array}{ccc}\omega^0\{A;B\} & \xrightarrow{\text{Hurewicz}} & K^0\{A;B\}\\ \text{Hurewicz}\big\downarrow & & \big\downarrow\text{ch (Chern character)}\\ H^0\{A;B\} & \xrightarrow{\iota} & \mathfrak{C}^0\{A;B\}\end{array}\tag{1.7}$$

The map ι is given by tensoring with $\mathbf{Q}$ and including the summand $H^0\{A;B\}\otimes\mathbf{Q}$ in $\mathfrak{C}^0\{A;B\}$, (1.6).

A $\mathbb{T}$-equivariant version of this square will be needed; we discuss this next. Suppose now that A and B are compact $\mathbb{T}$-ENR's with basepoint. The one-point-compactification of a (finite-dimensional) real $\mathbb{T}$-module V will be written with a superscript "+" as V^+; it is a sphere with basepoint at infinity[3]. For the purposes of this paper a *multiplicative* $\mathbb{T}$*-equivariant cohomology theory* $h_{\mathbb{T}}$ is a sequence of functors: $(A,B)\mapsto h^i_{\mathbb{T}}\{A;B\}$, indexed by $i\in\mathbf{Z}$, which is equipped with products, *composition* $\circ$: $h^j_{\mathbb{T}}\{B;C\}\otimes h^i_{\mathbb{T}}\{A;B\}\longrightarrow h^{j+i}_{\mathbb{T}}\{A;C\}$ and *smash product* $\wedge: h^i_{\mathbb{T}}\{A;B\}\otimes h^{i'}_{\mathbb{T}}\{A';B'\}\to h^{i+i'}_{\mathbb{T}}\{A\wedge A';B\wedge B'\}$, and satisfies the usual properties of a non-equivariant theory with, in addition, the *stability property*[4]: *for all real* $\mathbb{T}$*-modules* V*, the smash product with the identity on* V^+,

$$1\wedge: h^i_{\mathbb{T}}\{A;B\}\longrightarrow h^i_{\mathbb{T}}\{V^+\wedge A;V^+\wedge B\} \tag{1.8}$$

is an isomorphism. Stable cohomotopy $\omega_{\mathbb{T}}$ is the prototype of such a theory.

The familiar equivariant K-cohomology groups $\tilde{K}^i_{\mathbb{T}}(A)$ extend naturally to an equivariant cohomology theory $K_{\mathbb{T}}$. In practice it is often convenient to reduce the computation of $K^0_{\mathbb{T}}\{A;B\}$ to that of a standard K-cohomology group by using Bott periodicity and duality. (Indeed, this procedure can be used to give a (somewhat clumsy) constructive definition of $K^0_{\mathbb{T}}\{A;B\}$.)

Equivariant classical cohomology is to be *Borel cohomology*. Thus,

$$H^0_{\mathbb{T}}\{A;B\} = \text{“}\varprojlim_{P} H^0_{P/\mathbb{T}}\{(P\times A)/\mathbb{T};(P\times B)/\mathbb{T}\}\text{”}, \tag{1.9}$$

where the inverse limit is taken over the category of compact free $\mathbb{T}$-ENR's P and equivariant maps and $H^0_{P/\mathbb{T}}$ is cohomology over the base $P/\mathbb{T}$. Then, if X is a compact $\mathbb{T}$-ENR, $H^i_{\mathbb{T}}\{X_+;S^0\}$is the original Borel cohomology group $H^i((E\mathbb{T}\times X)/\mathbb{T})$. For the sake of logical rigour one should, of course, restrict the spaces P considered, say to the spheres $S(nE)$, $(n\geq 1)$, and then omit the inverted commas. Here and elsewhere $S(V)$ denotes the unit sphere in a vector space V; and nE means the direct sum $E\oplus\ldots\oplus E$ of n copies of E.

1.10 Remark. For a finite group G, the cohomology group $H^i_G\{A;B\}$ (of compact G-ENR's A and B) can be described in terms of the singular chain complexes $\tilde{S}_*(A)$, $\tilde{S}_*(B)$ of A and B: if $W_*\to\mathbf{Z}$ is a free $\mathbf{Z}[G]$-resolution of $\mathbf{Z}$, then $H^i_G\{A;B\}$ is the set of chain homotopy classes of $\mathbf{Z}[G]$-chain maps $W_*\otimes\tilde{S}_*(A)\to\tilde{S}_*(B)$ of degree i. Recent work on cyclic homology, which identifies the coefficient ring of $H_{\mathbb{T}}$

---

(3) A more common notation is S^V.

(4) Such theories, satisfying (1.8), are called "stable" in [14].

with the cyclic homology of $\mathbf{Z}$ ($HC_*^-(\mathbf{Z})$ in [17]), suggests that it should be possible to describe $H_{\mathbf{T}}^i\{A;B\}$ algebraically, too, as an homology group of a chain complex.

The theory $\mathfrak{C}_{\mathbf{T}}$ is defined similarly, with "$\mathfrak{C}$" replacing "H" in (1.9). It is worth noting that, whereas the limit in (1.9) is achieved for $P = S(nE)$ with n sufficiently large, this is not, in general, the case for the theory $\mathfrak{C}$. We have a direct product

$$\mathfrak{C}_{\mathbf{T}}^0\{A;B\} = \prod_{j\in\mathbf{Z}} H_{\mathbf{T}}^{2j}\{A;B\} \otimes \mathbf{Q}, \tag{1.11}$$

rather than a direct sum.

To complete the equivariant square (1.7) we have the Chern character $\mathrm{ch} : K_{\mathbf{T}} \to \mathfrak{C}_{\mathbf{T}}$, which is described, for example, in [2]. Thus, if X is a compact $\mathbf{T}$-ENR then $\mathrm{ch} : K_{\mathbf{T}}^0(X) \to \mathfrak{C}_{\mathbf{T}}^0(X)$ is defined by lifting from X to $P \times X$, identifying $K_{\mathbf{T}}^0(P \times X)$ with $K^0((P \times X)/\mathbf{T})$, and applying the non-equivariant Chern character:

$$\begin{aligned} K_{\mathbf{T}}^0(X) \longrightarrow \varprojlim K_{\mathbf{T}}^0(P \times X) &= \varprojlim K^0((P \times X)/\mathbf{T}) \\ &\longrightarrow \varprojlim \mathfrak{C}^0((P \times X)/\mathbf{T}) = \mathfrak{C}_{\mathbf{T}}^0(X). \end{aligned} \tag{1.12}$$

Next we consider the *classifying space*, $\mathbf{E}\mathfrak{F}$, *of the family* $\mathfrak{F}$ *of finite subgroups of* $\mathbf{T}$; see tom Dieck [14]. As a concrete realization of $\mathbf{E}\mathfrak{F}$ we can take the union, with the weak topology, of the spheres $S(\bigoplus_{1\le k\le n}(nE^k))$ for $n \ge 1$ (with nE^k included as the first n factors in $(n+1)E^k$). More loosely, we shall think of $\mathbf{E}\mathfrak{F}$ as the direct limit

$$\mathbf{E}\mathfrak{F} = \text{“}\varinjlim_F S(F)\text{”}, \tag{1.13}$$

where F runs over the category of (finite-dimensional) complex $\mathbf{T}$-modules (with invariant inner product) such that the fixed subspace $F^{\mathbf{T}}$ is trivial and the maps are isometric linear embeddings.

Now if X is a $\mathbf{T}$-ENR with fixed-point-free circle action, $X^{\mathbf{T}} = \emptyset$, there is a $\mathbf{T}$-map $X \to \mathbf{E}\mathfrak{F}$ and it is unique up to equivariant homotopy. Indeed, there is an embedding of X into some $\mathbf{T}$-module V, which we can write as $V^{\mathbf{T}} \oplus F$ with F as above. Projecting onto the second factor we obtain a map $X \to F - \{0\} \to S(F)$. The composition of two such maps $X \to S(F)$ and $X \to S(F')$ with the respective inclusions of $S(F)$ and $S(F')$ in $S(F \oplus F')$ are, of course, homotopic.

For any multiplicative $\mathbf{T}$-equivariant cohomology theory $h_{\mathbf{T}}$ defined on the category of compact $\mathbf{T}$-ENR's, we set

$$h_{\mathbf{T}}^i\{A; B \wedge (\mathbf{E}\mathfrak{F})_+\} = \varinjlim_F h_{\mathbf{T}}^i\{A; B \wedge S(F)_+\}, \tag{1.14}$$

where the limit is taken over complex representations F with $F^{\mathbf{T}} = 0$ as in (1.13). The invariants appearing in the title will lie in groups of this type. These groups are often most easily calculated by using the localization exact sequence, which we now recall. (The textbook [14] is again a good reference.)

For each $\mathbb{T}$-module F with $F^{\mathbb{T}} = 0$ we have a Gysin cofibration sequence:

$$S(F)_+ \hookrightarrow D(F)_+ \to D(F)/S(F) = F^+,$$

where $D(F)$ is the unit disc in F, or equivalently:

$$S(F)_+ \to S^0 \to S^0 * S(F) = F^+. \tag{1.15}$$

This gives rise to a long exact sequence:

$$\ldots \longrightarrow h^i_{\mathbb{T}}\{A; B \wedge S(F)_+\} \longrightarrow h^i_{\mathbb{T}}\{A; B\} \longrightarrow h^i_{\mathbb{T}}\{A; B \wedge F^+\} \longrightarrow \ldots$$

in cohomology. Forming the direct limit over the representations F, we obtain the *localization exact sequence*:

$$\ldots \longrightarrow h^i_{\mathbb{T}}\{A; B \wedge (\mathbf{E}\mathfrak{F})_+\} \longrightarrow h^i_{\mathbb{T}}\{A; B\} \longrightarrow \check{h}^i_{\mathbb{T}}\{A; B\} \longrightarrow \ldots, \tag{1.16}$$

in which we have written

$$\check{h}^i_{\mathbb{T}}\{A; B\} = \varinjlim_F h^i_{\mathbb{T}}\{A; B \wedge F^+\}. \tag{1.17}$$

In other words, $\check{h}_{\mathbb{T}}$ is obtained by inverting the Euler classes of all the representations E^k for $k \neq 0$. By construction, $\check{h}_{\mathbb{T}}$ is itself a multiplicative $\mathbb{T}$-equivariant cohomology theory. (The multiplication comes from the identification: $F^+ \wedge F'^+ \to (F \oplus F')^+$.) The key result of the localization theory is that $\check{h}^i_{\mathbb{T}}\{A; B\}$ depends only on the fixed subspaces $A^{\mathbb{T}}$ and $B^{\mathbb{T}}$.

1.18 Proposition. *The restriction and inclusion maps give isomorphisms:*

$$\check{h}^i_{\mathbb{T}}\{A; B\} \xrightarrow{\cong} \check{h}^i_{\mathbb{T}}\{A^{\mathbb{T}}; B\} \xleftarrow{\cong} \check{h}^i_{\mathbb{T}}\{A^{\mathbb{T}}; B^{\mathbb{T}}\}.$$

The sequence (1.16) thus, in some sense, resolves the equivariant theory into fixed-point-free and fixed components. A proof of the localization theorem for a general group can be found in [13] or [14]. It is instructive to specialize to the circle.

Outline proof of 1.18. The inclusion $A^{\mathbb{T}} \hookrightarrow A$ defines an element of $\check{\omega}^0_{\mathbb{T}}\{A^{\mathbb{T}}; A\}$. We shall write down an inverse in $\check{\omega}^0_{\mathbb{T}}\{A; A^{\mathbb{T}}\}$. Composition with this $\check{\omega}_{\mathbb{T}}$-map then defines an inverse $\check{h}^i_{\mathbb{T}}\{A^{\mathbb{T}}; B\} \to \check{h}^i_{\mathbb{T}}\{A; B\}$ to the restriction, and this establishes the first isomorphism. The second is similar.

Since A is a $\mathbb{T}$-ENR, we have an open set U in a $\mathbb{T}$-module V and $\mathbb{T}$-maps $i : A \to U$, $r : U \to A$ such that $r \circ i = 1$. To simplify notation we may assume that i is the inclusion of a subspace. Fix an invariant inner product on V and write V as an orthogonal sum $V^{\mathbf{T}} \oplus F$. By the compactness of A there exists $d > 0$ such that $\| v - a \| > d$ for all $v \in V - U$, $a \in A$.

We construct two homotopies $f_t, g_t : A \to A \wedge (D(F)/S(F)) = A \wedge F^+$, $t \in [0,1]$, with $f_1 = g_0$ as follows. Set

$$f_t(x,y) = \begin{cases} [(x,y), \frac{t}{d}y], & \text{if} \quad t \mid y \mid \leq d \\ * & \text{if} \quad t \mid y \mid \geq d \end{cases}$$

and

$$g_t(x,y) = \begin{cases} [r(x,(1-t)y), \frac{1}{d}y], & \text{if} \quad \mid y \mid \leq d \\ * & \text{if} \quad \mid y \mid \geq d \end{cases}$$

for $(x,y) \in A \subseteq V^{\mathbf{T}} \oplus F$. Then g_1 maps into $A^{\mathbf{T}} \wedge F^+$ and so defines an element of $\breve{\omega}^0_{\mathbf{T}}\{A; A^{\mathbf{T}}\}$. This is easily seen to be the required inverse of the inclusion $A^{\mathbf{T}} \to A$, because $f_0(a) = [a, 0]$ for $a \in A$ and $g_1(a) = [a, 0]$ for $a \in A^{\mathbf{Z}/2}$. ∎

If V and F are $\mathbf{T}$-modules with $F^{\mathbf{T}} = 0$, a $\mathbf{T}$-map $V^+ \wedge A^+ \to V^+ \wedge B^+ \wedge F^+$ restricts on fixed-points to a map $(V^{\mathbf{T}})^+ \wedge A^{\mathbf{T}} \to (V^{\mathbf{T}})^+ \wedge B^{\mathbf{T}}$. This construction defines the *fixed-point map*

$$\rho : \breve{\omega}^i_{\mathbf{T}}\{A; B\} \longrightarrow \omega^i\{A^{\mathbf{T}}; B^{\mathbf{T}}\}. \tag{1.19}$$

The result 1.18 leads immediately to the identification of $\breve{\omega}_{\mathbf{T}}$.

1.20 Proposition. *The fixed point map ρ, (1.19), is an isomorphism.*

Outline proof. By 1.18 we may assume that $\mathbf{T}$ acts trivially on A. Then ρ is clearly a split surjection (because any map is equivariant with respect to the trivial group-action), and it suffices to show that $\ker \rho = 0$. But a map $V^+ \wedge A^+ \to V^+ \wedge B^+$ and its restriction $(V^{\mathbf{T}})^+ \wedge A^+ \to V^+ \wedge B^+ = (V^{\mathbf{T}})^+ \wedge B^+ \wedge F^+$, where $V = V^{\mathbf{T}} \oplus F$, define the same element of $\breve{\omega}^0_{\mathbf{T}}\{A; B\}$. The injectivity of ρ follows readily. ∎

Finally, we record the computation in Borel cohomology of $\breve{H}_{\mathbf{T}}$.

1.21 Proposition. *There is an equivalence of multiplicative cohomology theories*

$$\breve{H}^i_{\mathbf{T}}\{A; B\} \xrightarrow{\cong} \mathfrak{C}^i\{A^{\mathbf{T}}; B^{\mathbf{T}}\}.$$

1.22 Remark. At the opposite extreme to the fixed-point-free actions we have the free actions, with classifying space $\mathbf{ET}$ instead of $\mathbf{E}\mathfrak{F}$. The corresponding localization procedure produces the *Tate cohomology theories*

$$\hat{h}^i_{\mathbf{T}}\{A; B\} = h^i_{\mathbf{T}}\{A; B \wedge (S^0 * \mathbf{ET})\} = \varinjlim_F h^i_{\mathbf{T}}\{A; B \wedge F^+\}, \tag{1.23}$$

where the limit this time is over the representations F for which $\mathbb{T}$ acts *freely* on $S(F)$. The localization exact sequence corresponding to (1.16) is the usual *Tate exact sequence* relating group homology, group cohomology and Tate cohomology[5]. In cyclic homology the emphasis seems to have been on $E\mathbb{T}$ and the Tate group $\hat{H}_{\mathbb{T}}$ rather than on $E\mathfrak{F}$ and $\check{H}_{\mathbb{T}}$. Of course, they agree if one is working with rational or complex coefficients. Perhaps in applications to analysis on loop manifolds the fixed-point-free theory will turn out to be the more natural.

Aside on $\mathbb{Z}/2$-homotopy. Much of the $\mathbb{T}$-equivariant homotopy theory described above parallels, with somewhat greater complexity, the $\mathbb{Z}/2$-equivariant theory recounted, for example in [9]. It is reasonable to look for analogues in the $\mathbb{T}$-theory of other features of the $\mathbb{Z}/2$-theory. For $\mathbb{Z}/2$, the fixed-point-free and free classifying spaces of course coincide. The $\mathbb{Z}/2$-Tate construction provides the natural home for various brands of Chern character. Let X be a compact ENR with trivial $\mathbb{Z}/2$-action. Corresponding to 1.20 and 1.21 we have: $\hat{\omega}^i_{\mathbb{Z}/2}(X) = \omega^i(X)$, and $\hat{H}^i_{\mathbb{Z}/2}(X) = \bigoplus_{j\in\mathbb{Z}} H^{i+2j}(X;\mathbb{F}_2)$. (Bredon, instead of Borel, cohomology gives a connective version.) For connective complex K-theory, $k_{\mathbb{Z}/2}$, the work of Adams, [1], on the $\mathbb{Z}_{(2)}$-Chern character can be reformulated as the computation of the Tate theory: $\hat{k}^i_{\mathbb{Z}/2}(X)_{(2)} = \bigoplus_{j\geq 0} H^{i+2j}(X;\mathbb{Z}_{(2)})$. For the K-theory of finite fields of odd characteristic one obtains the Witt cohomology theory.

2 Some computations

Let N be a complex $\mathbb{T}$-module of complex dimension r. In this section we shall examine the basic square:

$$\begin{array}{ccc} \omega_{\mathbb{T}}^{-1}\{N^+;(E\mathfrak{F})_+\} & \xrightarrow{\text{Hurewicz}} & K_{\mathbb{T}}^{-1}\{N^+;(E\mathfrak{F})_+\} \\ \Big\downarrow{\scriptstyle\text{Hurewicz}} & & \Big\downarrow{\scriptstyle\text{ch}} \\ H_{\mathbb{T}}^{-1}\{N^+;(E\mathfrak{F})_+\} & \xrightarrow{\iota} & \mathfrak{C}_{\mathbb{T}}^{-1}\{N^+;(E\mathfrak{F})_+\}. \end{array} \tag{2.1}$$

We start with the K-theory and write $R = K^0_{\mathbb{T}}(*) = K_0^{\mathbb{T}}(*)$ for the *complex representation ring* of $\mathbb{T}$: $R = \mathbb{Z}[z, z^{-1}]$, where $z^k = [E^k]$. Let S be the multiplicatively closed subset of R generated by the K-theory Euler classes $1 - z^n$ $(n \neq 0)$ of the representations E^n. One easily verifies that the segment

$$K_0^{\mathbb{T}}(*) \longrightarrow \check{K}_0^{\mathbb{T}}(*) \longrightarrow K_{-1}^{\mathbb{T}}(E\mathfrak{F})$$

of the localization exact sequence, (1.16), is given by the short exact sequence

[5] From this point of view the classical notation "H_*^G" for the homology of a finite group G is unhappy: it is incompatible with the terminology (1.3), and we must talk instead about $H_*^G(EG)$. There are similar problems with cyclic homology (rather than cohomology): $HC_{i-1}(\mathbb{Z})$, [17], is $H_i^{\mathbb{T}}(E\mathbb{T})$ here.

$$R \longrightarrow S^{-1}R \longrightarrow S^{-1}R/R. \tag{2.2}$$

Elements of the localization $S^{-1}R$ can be thought of as rational functions $f \in \mathbf{C}(z)$ of the form $f = p/q$ with $p \in \mathbf{Z}[z, z^{-1}]$ and q a product of factors of the form $1 - z^n$ $(n \neq 0)$. Then each class in $S^{-1}R/R$ can be described by its unique representative $f \in S^{-1}R$ which is holomorphic at 0 and ∞ and vanishes at ∞: $f(\infty) = 0$. The group $S^{-1}R/R$ has a direct sum decomposition:

$$S^{-1}R/R = \bigoplus_{n\geq 1} L_n \tag{2.3}$$

into components L_n (corresponding to the finite subgroups of $\mathbf{T}$) given by the functions whose only poles (in $\mathbf{C}$) are at primitive nth roots of unity.

We can now use the Bott isomorphism

$$\lambda_{\mathbf{C}} \cdot \lambda_N : K^{\mathbf{T}}_{-1}(\mathbf{E}\mathfrak{F}) = K^1_{\mathbf{T}}\{S^0; (\mathbf{E}\mathfrak{F})_+\} \to K^{-1}_{\mathbf{T}}\{N^+; (\mathbf{E}\mathfrak{F})_+\}$$

(as in [3] (2.7)) to make the identification:

$$K^{-1}_{\mathbf{T}}\{N^+; (\mathbf{E}\mathfrak{F})_+)\} = S^{-1}R/R. \tag{2.4}$$

The coefficient group $\mathfrak{C}^0_{\mathbf{T}}(*)$ is the formal power series ring $A = \mathbf{Q}[[t]]$ where $e^{-t} = \mathrm{ch}(z)$. Elements of the quotient field $K = \mathbf{Q}((t))$ will be written in the form $\sum_{i\geq -N} a_i t^i$ (that is, $t^{-N}(\sum_{i\geq 0} a_{i-N} t^i)$). The localization sequence

$$\mathfrak{C}^{\mathbf{T}}_0(*) \longrightarrow \check{\mathfrak{C}}^{\mathbf{T}}_0(*) \longrightarrow \mathfrak{C}^{\mathbf{T}}_{-1}(\mathbf{E}\mathfrak{F})$$

is then computed as

$$A \longrightarrow K \longrightarrow K/A, \tag{2.5}$$

and the Chern character from (2.2) to (2.5) is given at each level by $z \mapsto e^{-t}$.

The *Todd class*(6), $\mathcal{T}(N) \in A$, of N is defined as follows. We set

$$\mathcal{T}(E^k) = kt/(1 - e^{-kt}) \quad \text{if} \quad k \neq 0, \qquad \mathcal{T}(\mathbf{C}) = 1, \tag{2.6}$$

and extend $\mathcal{T}$ to a homomorphism: $[N] \mapsto \mathcal{T}(N)$ from R to the group of units in A.

The element $\mathcal{T}(N) \cdot \mathrm{ch}(\lambda_{\mathbf{C}} \cdot \lambda_N)$ is now a classical cohomology Thom class ([4: 2]), which we use to write

$$\mathfrak{C}^{-1}_{\mathbf{T}}\{N^+; (\mathbf{E}\mathfrak{F})_+\} = K/A. \tag{2.7}$$

(6) Depending on one's preferred sign convention, it may be the *dual* Todd class.

So the Chern character ch : $K_{\mathbf{T}}^{-1}\{N^+;(\mathbf{E}\mathfrak{F})_+\} \to \mathfrak{C}_{\mathbf{T}}^{-1}\{N^+;(\mathbf{E}\mathfrak{F})_+\}$ is given by:

$$f\,(\mathrm{mod}\,R) \in S^{-1}R/R \mapsto T(N)^{-1}\cdot f(e^{-t})\,(\mathrm{mod}\,A) \in K/A. \tag{2.8}$$

In integral cohomology we have: $H_{2i}^{\mathbf{T}}(*) = 0$, $\check{H}_{2i}^{\mathbf{T}}(*) \cong \mathbf{Q}$, $H_{2i-1}^{\mathbf{T}}(\mathbf{E}\mathfrak{F}) \cong \mathbf{Q}$, if $i > 0$, and $H_{2i}^{\mathbf{T}}(*) \cong \mathbf{Z}$, $\check{H}_{2i}^{\mathbf{T}}(*) \cong \mathbf{Q}$, $H_{2i-1}^{\mathbf{T}}(\mathbf{E}\mathfrak{F}) \cong \mathbf{Q}/\mathbf{Z}$, if $i \le 0$. Hence $H_{\mathbf{T}}^{-1}\{N^+;(\mathbf{E}\mathfrak{F})_+\}$ is isomorphic to $\mathbf{Q}$, by the Thom isomorphism; we prescribe a generator by giving the transformation $\iota : H_{\mathbf{T}}^{-1}\{N^+;(\mathbf{E}\mathfrak{F})_+\} \to \mathfrak{C}_{\mathbf{T}}^{-1}\{N^+;(\mathbf{E}\mathfrak{F})_+\}$ as:

$$a \in \mathbf{Q} \mapsto at^{-r-1}(\mathrm{mod}\,A) \in K/A. \tag{2.9}$$

Stable homotopy is, of course, more difficult. We look first at the case $N = 0$. Recall (from [14] say) that the group $\omega_1^{\mathbf{T}}(*)$ splits as a direct sum of components indexed by the (conjugacy classes of) subgroups of $\mathbf{T}$. The localization sequence

$$\omega_1^{\mathbf{T}}(\mathbf{E}\mathfrak{F}) \longrightarrow \omega_1^{\mathbf{T}}(*) \rightleftarrows \check{\omega}_1^{\mathbf{T}}(*) = \omega_1(*)$$

is split, and $\omega_1^{\mathbf{T}}(\mathbf{E}\mathfrak{F})$ is included in $\omega_1^{\mathbf{T}}(*)$ as the sum of the components corresponding to finite isotropy groups. We have

$$\omega_1^{\mathbf{T}}(\mathbf{E}\mathfrak{F}) = \bigoplus_{n\ge 1} \mathbf{Z}\sigma_n, \tag{2.10}$$

where the generators σ_n of the infinite cyclic summands are described as follows. The 1-dimensional $\mathbf{T}$-manifold $S(E^n)$, for $n \ge 1$, has a natural equivariant framing given by a trivialization $S(E^n)\times\mathbf{R} \to \tau S(E^n)$ of the tangent bundle. (See (3.1); in this case $\hat{\tau}$ is zero.) The class σ_n is obtained by applying the Pontrjagin-Thom construction to the fixed-point-free framed manifold $S(E^n)$ with its classifying map $S(E^n) \to (\mathbf{E}\mathfrak{F})_+$. It is elementary to compute its Hurewicz image in K-theory or cohomology (at least up to sign convention).

2.11 Lemma. *The Hurewicz maps*

$$\omega_1^{\mathbf{T}}(\mathbf{E}\mathfrak{F}) \to K_1^{\mathbf{T}}(\mathbf{E}\mathfrak{F}) = S^{-1}R/R \quad \textit{and} \quad \omega_1^{\mathbf{T}}(\mathbf{E}\mathfrak{F}) \to H_1^{\mathbf{T}}(\mathbf{E}\mathfrak{F}) = \mathbf{Q}$$

send the generator σ_n to $1/(1-z^n)$ and $1/n$ respectively.

The case $N = E$ can also be treated completely by considering the restriction map from N to the zero subspace. Write $i : 0 \to N$ for the inclusion. In the Gysin sequence

$$\ldots \to \omega_{\mathbf{T}}^{-1}\{E^+;(\mathbf{E}\mathfrak{F})_+\} \xrightarrow{i^*} \omega_{\mathbf{T}}^{-1}\{0^+;(\mathbf{E}\mathfrak{F})_+\} \to \omega_{\mathbf{T}}^{-1}\{S(E)^+;(\mathbf{E}\mathfrak{F})_+\} \to \ldots \tag{2.12}$$

the term $\omega_{\mathbf{T}}^{-1}\{S(E)^+;(\mathbf{E}\mathfrak{F})_+\}$ can be rewritten as $\omega_{\mathbf{T}}^{-1}\{\mathbf{T}^+;(\mathbf{E}\mathfrak{F})_+\}$ and thus identified with the non-equivariant group $\omega^{-1}\{S^0;(\mathbf{E}\mathfrak{F})_+\} = \mathbf{Z}/2\cdot\eta$. The map $\omega_1^{\mathbf{T}}(\mathbf{E}\mathfrak{F}) = \omega_{\mathbf{T}}^{-1}\{S^0;(\mathbf{E}\mathfrak{F})_+\} \to \omega^{-1}\{S^0;(\mathbf{E}\mathfrak{F})_+\} = \omega_1(*)$ in (2.12) simply forgets the $\mathbf{T}$-action; it sends σ_n to η and is certainly surjective. Multiplying by η we conclude, because $\omega_2(*) = \mathbf{Z}/2\cdot\eta^2$, that i^* in (2.12) is injective. This effects the computation of $\omega_{\mathbf{T}}^{-1}\{E^+;(\mathbf{E}\mathfrak{F})_+\}$.

2.13 **Lemma.** *The restriction map*

$$i^*: \omega_{\mathsf{T}}^{-1}\{E^+; (\mathbf{E}\mathfrak{F})_+\} \longrightarrow \omega_{\mathsf{T}}^{-1}\{0^+; (\mathbf{E}\mathfrak{F})_+\}$$

is injective, with image $\{\sum a_n\sigma_n \mid \sum a_n \text{ is even}\} \subseteq \bigoplus_{n\geq 1} \mathbf{Z}\sigma_n$. ∎

Similar arguments can be used when $N = E^k$, $(k \geq 1)$. The circle $S(E^k)$ can be identified with the homogeneous space $\mathsf{T}/\mathsf{T}(k)$. So we can write $\omega_{\mathsf{T}}^{-1}\{S(E^k)_+; (\mathbf{E}\mathfrak{F})_+\}$ as $\omega_{\mathsf{T}(k)}^{-1}\{S^0; (\mathbf{E}\mathfrak{F})_+\} = \omega_{\mathsf{T}(k)}^{-1}(*)$. Recollect that this group, too, splits as a direct sum

$$\omega_1^{\mathsf{T}(k)}(*) = \bigoplus_{l|k} \omega_1(B(\mathsf{T}(k)/\mathsf{T}(l))),\ (l \geq 1), \tag{2.14}$$

which can be written down explicitly by using the isomorphism

$$\omega_1(B(\mathbf{Z}/m)) \xrightarrow{\cong} \mathbf{Z}/2 \oplus \mathbf{Z}/m \quad (\text{for } m \geq 1) \tag{2.15}$$

(given by projection to a point: $\omega_1(B(\mathbf{Z}/m)) \to \omega_1(*) = \mathbf{Z}/2$ and the Hurewicz map: $\omega_1(B(\mathbf{Z}/m)) \to H_1(B(\mathbf{Z}/m)) = \mathbf{Z}/m$). From the Gysin sequence for $S(E^k)$ we obtain the following description of the image of i^*.

2.16 **Lemma.** *The element* $\sum a_n\sigma_n$ *lies in the image of*

$$i^*: \omega_{\mathsf{T}}^{-1}\{(E^k)^+; (\mathbf{E}\mathfrak{F})_+\} \longrightarrow \omega_{\mathsf{T}}^{-1}\{0^+; (\mathbf{E}\mathfrak{F})_+\}$$

if and only if for each divisor l *of* k, (i) $\sum_{(n,k)=l} a_n$ *is even, and in the ring of integers* (mod k/l) *we have* (ii) $\sum_{(n,k)=l} a_n \cdot (n/l)^{-1} = 0$. ∎

In particular, the cokernel of i^* is finite, as is its kernel, which is a quotient of the finite group $\omega_2^{\mathsf{T}(k)}(*)$. An inductive proof establishes the corresponding result for any module N with $N^{\mathsf{T}} = 0$.

2.17 **Lemma.** *Suppose that* $N^{\mathsf{T}} = 0$. *Then the restriction map*

$$i^*: \omega_{\mathsf{T}}^{-1}\{N^+; (\mathbf{E}\mathfrak{F})_+\} \longrightarrow \omega_{\mathsf{T}}^{-1}\{0^+; (\mathbf{E}\mathfrak{F})_+\}$$

has finite kernel and cokernel.

Proof. The argument above generalizes to give an exact Gysin sequence:

$$\omega_{\mathsf{T}(k)}^{-2}\{N^+; S^0\} \longrightarrow \omega_{\mathsf{T}}^{-1}\{(N \oplus E^k)^+; (\mathbf{E}\mathfrak{F})_+\} \longrightarrow \omega_{\mathsf{T}}^{-1}\{N^+; (\mathbf{E}\mathfrak{F})_+\} \longrightarrow \omega_{\mathsf{T}(k)}^{-1}\{N^+; S^0\},$$

for $k \geq 1$. So it will suffice to show that

$$\omega_{\mathsf{T}(k)}^{-i}\{N^+; S^0\} \quad \text{is finite for } i > 0. \tag{2.18}$$

But N^+ has a $\mathsf{T}(k)$-equivariant cell decomposition (and it is not hard to write down an explicit decomposition in this case) into cells of the form $(\mathsf{T}(k)/\mathsf{T}(l)) \times \mathbf{R}^d$ (with $d \leq 2r$). The assertion (2.18) follows by a five-lemma argument from the finiteness of $\omega_j^{\mathsf{T}(l)}(*)$ for $j > 0$. ∎

We shall need a more precise lifting.

2.19 Lemma. *In 2.17 some (non-zero) multiple of σ_n, $n \geq 1$, has a lifting lying in the image of*

$$\omega_{\mathbf{T}}^{-1}\{N^+; S((r+1)E^n)_+\} \longrightarrow \omega_{\mathbf{T}}^{-1}\{N^+; (\mathbf{E}\mathfrak{F})_+\}.$$

Proof. The method refines that of 2.17. We show that the group $\omega_{\mathbf{T}(k)}^{-1}\{N^+; S(qE^n)_+\}$ is finite for $q > r+1$. Now $\tilde{\omega}_j^{\mathbf{T}(l)}((q\,E^n)^+)$ is a direct sum of non-equivariant groups, [14: II.7], and is thus seen to be finite if $0 < j < 2q$. An induction over cells, as above, then shows that the first term in the Gysin sequence

$$\omega_{\mathbf{T}(k)}^{-2}\{N^+; (qE^n)^+\} \longrightarrow \omega_{\mathbf{T}(k)}^{-1}\{N^+; S(qE^n)_+\} \longrightarrow \omega_{\mathbf{T}(k)}^{-1}\{N^+; S^0\}$$

is finite if $q > r+1$. The third term is finite by (2.18), and this finishes the proof. ∎

According to 2.17 the Hurewicz image in $H_{\mathbf{T}}^{-1}\{N^+; (\mathbf{E}\mathfrak{F})_+\}$ or $K_{\mathbf{T}}^{-1}\{N^+; (\mathbf{E}\mathfrak{F})_+\}$, both torsion-free, of a class $x \in \omega_{\mathbf{T}}^{-1}\{N^+; (\mathbf{E}\mathfrak{F})_+\}$ must be determined by $i^*(x)$. The calculation in cohomology is easy.

2.20 Lemma. *Suppose that $N^{\mathbf{T}} = 0$: $N = \bigoplus_{1\leq i\leq r} E^{k_i}$. Let the element $x \in \omega_{\mathbf{T}}^{-1}\{N^+; (\mathbf{E}\mathfrak{F})_+\}$ restrict to $i^*(x) = \sum a_n\sigma_n \in \omega_{\mathbf{T}}^{-1}\{0^+; (\mathbf{E}\mathfrak{F})_+\}$. Then the Hurewicz image of x in $H_{\mathbf{T}}^{-1}\{N^+; (\mathbf{E}\mathfrak{F})_+\} = \mathbf{Q}$ is*

$$(k_1\,k_2 \ldots k_r)^{-1} \cdot (\textstyle\sum a_n/n).$$

∎

For the K-theory, observe that a rational function $g \in L_m$, (2.3), is determined by $g(\zeta_m e^{-t}) \in \mathbf{Q}(\zeta_m)((t))/\mathbf{Q}(\zeta_m)[[t]]$, where ζ_m is a primitive mth root of unity, say $\zeta_m = e^{2\pi i/m}$.

2.21 Lemma. *Write $f = \sum f_m \in S^{-1}R/R = \bigoplus L_m$ for the K-theory Hurewicz image of the class x in 2.20. Then f_m is given by:*

$$f_m(\zeta_m e^{-t}) = \Bigg(\sum_{n\equiv 0 (\mathrm{mod}\ m)} a_n/n\Bigg)\cdot t^{-1} \cdot \prod_{1\leq j\leq r} (1-\zeta_m^{k_j} e^{-k_j t})^{-1} (\mathrm{mod}\ \mathbf{Q}(\zeta_m)[[t]]).$$

Results of this type can be found in [2:9].

Proof of 2.21. The restriction map

$$K_{\mathbf{T}}^{-1}\{(N \oplus E^k)^+; (\mathbf{E}\mathfrak{F})_+\} \longrightarrow K_{\mathbf{T}}^{-1}\{N^+; (\mathbf{E}\mathfrak{F})_+\} \tag{2.22}$$

(for $k \geq 1$) is given by multiplication by the K-theory Euler class $1 - z^k$: $S^{-1}R/R \to S^{-1}R/R$.

Since $1-\zeta_m^k e^{-kt}$ is invertible in $\mathbb{Q}(\zeta_m)[[t]]$ if k is not divisible by m, we can reduce at once to the case in which m divides each k_j $(1 \le j \le r)$.

It is enough to treat the case in which $i^*(x)$ is a multiple of σ_n for some $n \ge 1$. By (2.19), we then have $(1-z^n)^{r+1} f(z) \in R$, and $f_m = 0$ unless $n \equiv 0 (\operatorname{mod} m)$. So assume further that m divides n.

By using the mth power map $\mathbf{T} \to \mathbf{T}$ we can then reduce to the case $m = 1$. But $f(e^{-t})$ is determined by the cohomology calculations (2.8),(2.9),2.20, from the commutative square (2.1). ∎

In the special case $r = 1$, Lemma 2.21 easily yields an explicit formula for the rational function f.

2.23 Corollary. *Let $x \in \omega_{\mathbf{T}}^{-1}\{(E^k)^+; (\mathbf{E}\mathfrak{F})_+\}$, where $k \ge 1$, be a class restricting to $\sum a_n \sigma_n \in \omega_{\mathbf{T}}^{-1}\{0^+; (\mathbf{E}\mathfrak{F})_+\}$. Then the K-theory Hurewicz image of x in $K_{\mathbf{T}}^{-1}\{(E^k)^+; (\mathbf{E}\mathfrak{F})_+\} = S^{-1}R/R$ is given by $\sum a_n \psi_{n,k}(z)$, where $\psi_{n,k}(z) \in \mathbb{Q}(z)$ is the following rational function with poles at the nth roots of unity $\zeta : \zeta^n = 1$.*

$$\psi_{n,k}(z) = \frac{1}{nk} \sum_{\substack{\zeta^n=1 \\ \zeta^k=1}} \left(\frac{\zeta^2}{(\zeta - z)^2} + (\frac{k}{2} - 1)\frac{\zeta}{\zeta - z} \right) - \frac{1}{n} \sum_{\substack{\zeta^n=1 \\ \zeta^k \ne 1}} \frac{1}{\zeta^k - 1} \cdot \frac{\zeta}{\zeta - z}.$$

∎

It is, perhaps, a little simpler to adapt the argument of 2.21. If $(n,k) = l$, then we can use the lth power map to see that

$$\psi_{n,k}(z) = \psi_{n/l,k/l}(z^l). \tag{2.24}$$

So assume that n and k are coprime. From 2.11 and the computation of the restriction map (2.22) we have: $(1-z^k)\psi_{n,k}(z) \equiv 1/(1-z^n) (\operatorname{mod} R)$. By 2.19 all the poles of $\psi_{n,k}$ are at nth roots of unity, and cohomology determines the behaviour at $z = 1$. So we can read off $\psi_{n,k}$ from the partial fraction expansion of $1/(1-z^k)(1-z^n)$. In this way we obtain the following alternative formula for $\psi_{n,k}$ when $(n,k) = 1$.

$$\psi_{n,k}(z) = \frac{1}{(1-z^k)(1-z^n)} + \frac{1}{k} \sum_{\substack{\zeta^k=1 \\ \zeta \ne 1}} \frac{1}{\zeta^n - 1} \cdot \frac{\zeta}{\zeta - z} - \frac{1}{2k} \cdot \frac{1}{1-z} \tag{2.25}$$

$$= \frac{1}{nk} \cdot \frac{1}{(1-z)^2} + \frac{1}{nk}(\frac{k}{2} - 1)\frac{1}{1-z} - \frac{1}{n} \sum_{\substack{\zeta^n=1 \\ \zeta \ne 1}} \frac{1}{\zeta^k - 1} \cdot \frac{\zeta}{\zeta - z}. \tag{2.26}$$

2.27 Remark. The restriction $N^{\mathbf{T}} = 0$ imposed in the latter part of this section is not an essential one. It is made only for the sake of a uniform, and reasonably brief, exposition. In the general case one has the splitting

$$\omega_{2q+1}^{\mathbf{T}}(\mathbf{E}\mathfrak{F}) = \bigoplus_{n \ge 1} \omega_{2q}(\mathbf{B}(\mathbf{T}/\mathbf{T}(n)))$$

(for $q = \dim_{\mathbf{C}} N^{\mathbf{T}}$). Each summand is $\omega_{2q}(\mathbf{BT})$ and modulo torsion is infinite cyclic. So we can again associate to each element of $\omega_{2q+1}^{\mathbf{T}}(\mathbf{E}\mathfrak{F})$ a sequence of integers (a_n). The Hurewicz image in $H_{2q+1}^{\mathbf{T}}(\mathbf{E}\mathfrak{F}) = \mathbf{Q}$ is given, with appropriate choices of sign, by $q! \sum a_n/n^{q+1}$.

3 Geometry

From now on, M will be a finite-dimensional closed (so compact) smooth $\mathbf{T}$-manifold with a fixed-point-free action. This circle action is generated by a nowhere-zero smooth vector field, s say. (To be precise we can define s_x at a point $x \in M$ to be the derivative at 0 of the map $\mathbf{R} \to M$: $t \mapsto e^{2\pi i t} \cdot x$.) Then, choosing an invariant Riemannian metric on M, we can split the tangent space $\tau_x M$ at each point $x \in M$ as a sum $\mathbf{R}\, s_x \oplus \hat{\tau}_x$ of the 1-dimensional subspace, spanned by s_x, tangent to the orbit through x and its orthogonal complement, normal to the orbit. This gives a global decomposition

$$\tau M = \mathbf{R} \oplus \hat{\tau}, \tag{3.1}$$

which is, up to homotopy, independent of the choice of metric.

More generally, if a Lie group G, with Lie algebra $\mathfrak{g}$ (on which G acts by the adjoint representation), acts on a smooth manifold X, we have a homomorphism:

$$X \times \mathfrak{g} \longrightarrow \tau X \tag{3.2}$$

of G-vector bundles over X. It is given on fibres at $x \in X$ by the derivative $\tau_1 G = \mathfrak{g} \to \tau_x X$ of the map $G \to X : g \mapsto gx$ at $1 \in G$.

In particular, if Γ is a discrete subgroup of G and we take X to be the homogeneous space G/Γ (with G acting by left translation), we obtain a G-equivariant framing:

$$(G/\Gamma) \times \mathfrak{g} \longrightarrow \tau(G/\Gamma). \tag{3.3}$$

This leads us to an interesting class of examples, studied by Seade and Steer in [21] and [22], which will be used to illustrate the general results of Sections 4 and 5. Let Γ be a *co-compact Fuchsian group*, that is, a discrete subgroup of $\mathbf{PSL}(2, \mathbf{R})$ with compact quotient

$$M = \mathbf{PSL}(2, \mathbf{R})/\Gamma. \tag{3.4}$$

We identify $\mathbf{T}$ with the subgroup $\mathbf{PSO}(2) \leq \mathbf{PSL}(2, \mathbf{R})$ by

$$e^{2\pi i t} \mapsto \pm \begin{pmatrix} \cos \pi t & \sin \pi t \\ -\sin \pi t & \cos \pi t \end{pmatrix} \in \mathbf{SL}(2, \mathbf{R})/\{\pm 1\}.$$

Then $\mathbf{T}$ acts on M by left multiplication, without fixed points (because, if $z \in \mathbf{T}$ fixes $g\Gamma \in M$, then z lies in the discrete subgroup $g\Gamma g^{-1}$).

Identify the Lie algebra $\mathfrak{psl}(2,\mathbf{R})$ of $\mathbf{PSL}(2,\mathbf{R})$ with the Lie algebra $\mathfrak{sl}(2,\mathbf{R})$ of the double cover, and endow it with the invariant inner product: $\langle A, B\rangle = \mathrm{tr}(A\,B^*)/2\pi^2$. As $\mathbf{T}$-module, $\mathfrak{psl}(2,\mathbf{R})$ is then an orthogonal sum $\mathbf{R} \oplus E$ of components generated by unit vectors $\left(\begin{smallmatrix} 0 & \pi \\ -\pi & 0 \end{smallmatrix}\right)$ and $\left(\begin{smallmatrix} \pi & 0 \\ 0 & -\pi \end{smallmatrix}\right)$. By left translation, (3.3), we can trivialize τM as $M \times (\mathbf{R} \oplus E)$ and give M the corresponding metric. This splitting is then compatible with the decomposition (3.1) into summands tangential and normal to the orbits and produces a $\mathbf{T}$-equivariant trivialization:

$$M \times E \xrightarrow{\cong} \hat{\tau}. \tag{3.5}$$

4 The e-invariant

In this section we suppose that we are given, quite generally, a $\mathbf{T}$-equivariant trivialization: $M \times N \xrightarrow{\cong} \hat{\tau}$ (up to homotopy, with M and $\hat{\tau}$ as in (3.1)), where N, as in Section 2, is a complex $\mathbf{T}$-module of dimension r.

Using (3.1) we obtain an equivariant framing $\tau M \cong M \times (\mathbf{R} \oplus N)$ of M. This framed manifold together with the classifying map $M \to \mathbf{E}\mathfrak{F}$ represents, via the Pontrjagin-Thom construction, a stable cohomotopy class, x_M say, in $\omega_{\mathbf{T}}^{-1}\{N^+; (\mathbf{E}\mathfrak{F})_+\}$. Let $f_M \in S^{-1}R/R = K_{\mathbf{T}}^{-1}\{N^+; (\mathbf{E}\mathfrak{F})_+\}$ be its K-theory Hurewicz image.

On the other hand, if we forget the $\mathbf{T}$-action, the non-equivariantly framed manifold M ($\tau M \cong M \times \mathbf{R}^{2r+1}$) represents a class in $\omega_{2r+1}(*)$. Extending the fundamental work of Knapp, ([18], [19] and [20]), for free $\mathbf{T}$-actions, we shall show in this section that the complex e-invariant of the framed manifold M is determined by the K-theory invariant $f_M \in S^{-1}R/R$.

We shall think of the e-invariant in terms of stable cohomotopy and K-theory with $\mathbf{Q}/\mathbf{Z}$-coefficients, which we write as $\omega\mathbf{Q}/\mathbf{Z}$ and $K\mathbf{Q}/\mathbf{Z}$. Let us recall the definition. Suppose A and B are compact ENR's with basepoint and that $H^i\{A; B\} \otimes \mathbf{Q} = 0$, $H^{i-1}\{A; B\} \otimes \mathbf{Q} = 0$. Then the Bockstein homomorphism $\beta : (\omega\mathbf{Q}/\mathbf{Z})^{i-1}\{A; B\} \to \omega^i\{A; B\}$ is an isomorphism. Writing $\mathbf{d}$ throughout this section for the K-theory Hurewicz map, we define

$$\mathbf{e} : \ker\Big\{\mathbf{d}\colon \omega^i\{A;B\} \to K^i\{A;B\}\Big\} \to \mathrm{coker}\Big\{K^{i-1}\{A;B\} \to K^{i-1}\{A;B\} \otimes \mathbf{Q}\Big\}$$
$$\subseteq (K\mathbf{Q}/\mathbf{Z})^{i-1}\{A;B\} \tag{4.1}$$

by $\mathbf{e}(a) = \mathbf{d}(\beta^{-1}(a))$. For $A = B = S^0$ and $i = -2r-1 < 0$ this gives the classical e-invariant

$$\mathbf{e} : \omega_{2r+1}(*) \longrightarrow \mathbf{Q}/\mathbf{Z}. \tag{4.2}$$

4.3 Proposition. *The complex* **e**-*invariant in* $\mathbf{Q}/\mathbf{Z}$ *of the framed manifold* M *is equal to the constant term* (mod $\mathbf{Z}$) *in the formal power series* $\mathcal{T}(N)^{-1}\cdot f_M(e^{-t}) \in \mathbf{Q}((t))$.

In fact, it is no more difficult to prove a rather stronger result, and this is described next. For each $n \geq 1$, let $P(\mathbf{C}^n)^\nu$ denote the Thom space of the complex vector bundle $\nu = (S(nE)\times N)/\mathbf{T}$ over complex projective space $P(\mathbf{C}^n) = S(nE)/\mathbf{T}$. Lifting from a point to $S(nE)$ and applying the e-invariant (4.1) we obtain an equivariant **e**-invariant

$$\begin{aligned}\mathbf{e} : \tilde{\omega}_{\mathbf{T}}^{-1}(N^+) &\longrightarrow \varprojlim_n \tilde{\omega}_{\mathbf{T}}^{-1}(S(nE)_+ \wedge N^+) = \varprojlim_n \tilde{\omega}^{-1}(P(\mathbf{C}^n)^\nu) \\ &\longrightarrow \varprojlim_n (\tilde{K}\mathbf{Q}/\mathbf{Z})^{-2}(P(\mathbf{C}^n)^\nu). \end{aligned} \tag{4.4}$$

The completion of the representation ring $R = \mathbf{Z}[z, z^{-1}]$ with respect to the augmentation ideal will be written as $\hat{R} = \mathbf{Z}[[y]]$, $y = 1 - z$. Thus $K^0(P(\mathbf{C}^n)) = \hat{R}/(y^n)$. Then the last term in (4.4) can be identified, thanks to the Bott class $\lambda_{\mathbf{C}} \cdot \lambda_N$, with $\hat{R}\otimes \mathbf{Q}/\hat{R}$.

To state the main result of this section we require the map

$$\varepsilon_N : S^{-1}R \longrightarrow \hat{R}\otimes\mathbf{Q}$$

defined as follows. Let $\pi_+ : \mathbf{Q}((t)) \to \mathbf{Q}[[t]]$ project $\sum_{i\geq -N} a_i t^i$ to $\sum_{i\geq 0} a_i t^i$. Then, for $f(z) \in S^{-1}R$, $(\varepsilon_N f)(y) \in \mathbf{Q}[[y]] = \hat{R}\hat{\otimes}\mathbf{Q}$ is given by

$$\pi_+(\mathcal{T}(N)^{-1}\cdot f(e^{-t})) = \mathcal{T}(N)^{-1}\cdot(\varepsilon_N f)(1-e^{-t}) \in \mathbf{Q}[[t]]. \tag{4.5}$$

(The assignment $g(y) \in \mathbf{Q}[[y]] \mapsto g(1-e^{-t}) \in \mathbf{Q}[[t]]$ is evidently an isomorphism.) The restriction of ε_N to R is just the inclusion: $R \subseteq \hat{R} \subseteq \hat{R}\otimes\mathbf{Q}$. So we have an induced map $\varepsilon_N : S^{-1}R/R \longrightarrow \hat{R}\otimes\mathbf{Q}/\hat{R}$.

4.6 Proposition. *The following square commutes.*

$$\begin{array}{ccc} \omega_{\mathbf{T}}^{-1}\{N^+;(\mathbf{E}\mathfrak{F})_+\} & \longrightarrow & \tilde{\omega}_{\mathbf{T}}^{-1}(N^+) \\ \big\downarrow \mathbf{d} & & \big\downarrow \mathbf{e} \\ S^{-1}R/R & \xrightarrow{\varepsilon_N} & \hat{R}\otimes\mathbf{Q}/\hat{R}. \end{array}$$

The map **e** *is the the equivariant* **e**-*invariant, (4.4), and* ε_N *is defined by (4.5).*

The element of $\tilde{\omega}_{\mathbf{T}}^{-1}(N^+)$ represented by the equivariantly framed manifold M is just the restriction of the class $x_M \in \omega_{\mathbf{T}}^{-1}\{N^+;(\mathbf{E}\mathfrak{F})_+\}$. Its **e**-invariant is computed by 4.6 as $\varepsilon_N(f_M) \in \hat{R}\otimes\mathbf{Q}/\hat{R}$. Forgetting the $\mathbf{T}$-action we obtain 4.3.

Proof of 4.6. As usual we replace $\mathbf{E}\mathfrak{F}$ by $S(F)$ where F is a complex $\mathbf{T}$-module with $F^{\mathbf{T}} = 0$. We fix an integer $n \geq 1$ and work at the finite level:

$$\begin{aligned}\omega_{\mathbf{T}}^{-1}\{N^+;S(F)_+\} \longrightarrow \tilde{\omega}_{\mathbf{T}}^{-1}(N^+) \longrightarrow \tilde{\omega}_{\mathbf{T}}^{-1}(S(nE)_+\wedge N^+) = \tilde{\omega}^{-1}(P(\mathbf{C}^n)^\nu) \\ \longrightarrow (\tilde{K}\mathbf{Q}/\mathbf{Z})^{-2}(P(\mathbf{C}^n)^\nu).\end{aligned} \tag{4.7}$$

Put $m = n + \dim_{\mathbf{C}} F$. Now the e-invariant on $\omega_{\mathbf{T}}^{-1}\{N^+; S(F)_+\}$ factors through the composition:

$$\begin{aligned} &\omega_{\mathbf{T}}^{-1}\{N^+;S(F)_+\} \longrightarrow \omega_{\mathbf{T}}^{-1}\{S(mE)_+ \wedge N^+; S(F)_+\} \\ &\qquad \longrightarrow \omega_{\mathbf{T}}^{-1}\{S(mE)_+ \wedge N^+; S^0\} \longrightarrow \omega_{\mathbf{T}}^{-1}\{S(nE)_+ \wedge N^+; S^0\} \end{aligned} \tag{4.8}$$

induced by maps $S(mE) \to *$, $S(F) \to *$ and the inclusion $S(nE) \to S(mE)$. Using duality we can rewrite the second group in (4.8) as $\omega_{\mathbf{T}}^{-2}\{S(mE)_+ \wedge S(F)_+ \wedge N^+; F^+\}$ and the second arrow as the boundary homomorphism δ in a Gysin sequence. Let ϕ be the vector bundle $(S(mE) \times F)/\mathbf{T}$ over $P(\mathbf{C}^m)$ and $\pi : S(\phi) \to P(\mathbf{C}^m)$ the projection from the sphere-bundle to the base. Upon dividing out free $\mathbf{T}$-actions, (4.8) then becomes

$$\begin{aligned} &\omega_{\mathbf{T}}^{-1}\{N^+; S(F)_+\} \longrightarrow \tilde{\omega}^{-2}(S(\phi)^{\pi^*(\nu-\phi)}) \xrightarrow{\ \delta\ } \\ &\qquad\qquad \tilde{\omega}^{-1}(P(\mathbf{C}^m)^\nu) \xrightarrow{\ j^*\ } \tilde{\omega}^{-1}(P(\mathbf{C}^n)^\nu), \end{aligned} \tag{4.9}$$

where $j : P(\mathbf{C}^n) \to P(\mathbf{C}^m)$ is the inclusion.

At this stage it is convenient to introduce some temporary abbreviations:

$$\begin{aligned} P &= \omega^0\{P(\mathbf{C}^n)^\nu; P(\mathbf{C}^m)^{\nu-\phi}\}, \\ Q &= (\omega\mathbf{Q}/\mathbf{Z})^0\{P(\mathbf{C}^n)^\nu; P(\mathbf{C}^m)^{\nu-\phi}\}, \\ R &= (\omega\mathbf{Q}/\mathbf{Z})^0\{P(\mathbf{C}^n)^\nu; S(\phi)^{\pi^*(\nu-\phi)}\}, \\ S &= \omega^1\{P(\mathbf{C}^n)^\nu; S(\phi)^{\pi^*(\nu-\phi)}\}, \\ T &= \omega^0\{P(\mathbf{C}^n)^\nu; P(\mathbf{C}^m)^\nu\}. \end{aligned}$$

Then we have a commutative diagram

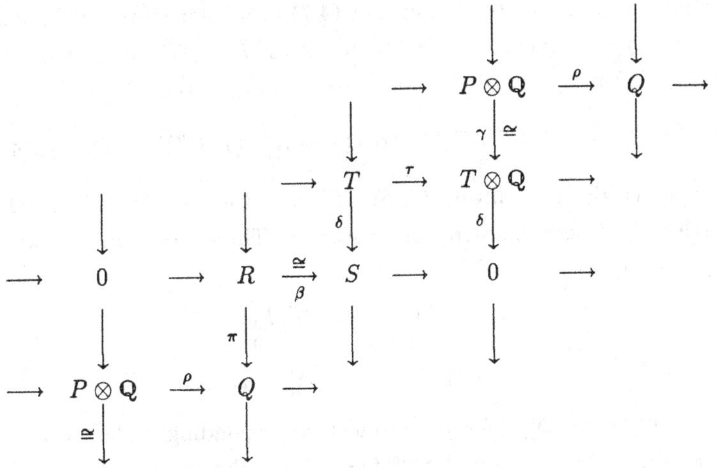

$$\begin{array}{ccccccccccc}
 & & & & & & & & \downarrow & & \downarrow & \\
 & & & & & & & \longrightarrow & P\otimes\mathbf{Q} & \xrightarrow{\rho} & Q & \longrightarrow \\
 & & & & & & \downarrow & & \gamma\downarrow\cong & & \downarrow & \\
 & & & & & \longrightarrow & T & \xrightarrow{\tau} & T\otimes\mathbf{Q} & \longrightarrow & & \\
 & \downarrow & & \downarrow & & & \delta\downarrow & & \delta\downarrow & & & \\
\longrightarrow & 0 & \longrightarrow & R & \xrightarrow[\beta]{\cong} & & S & \longrightarrow & 0 & \longrightarrow & & \\
 & \downarrow & & \pi\downarrow & & & \downarrow & & \downarrow & & & \\
\longrightarrow & P\otimes\mathbf{Q} & \xrightarrow{\rho} & Q & \longrightarrow & & & & & & & \\
 & \cong\downarrow & & \downarrow & & & & & & & &
\end{array} \tag{4.10}$$

in which the rows are Bockstein exact sequences and the columns Gysin exact sequences for the sphere-bundle $S(\phi)$, and where the rational groups are obtained by easy cohomology calculations using (1.5) and the condition: $m \geq n + \dim_{\mathbf{C}} F$.

Let $j_\# \in T$ denote the class given by the inclusion $P(\mathbf{C}^n)^\nu \to P(\mathbf{C}^m)^\nu$. Then $\delta(j_\#) \in S$ is a stable map which induces the composition $j^* \cdot \delta$ in (4.9). From (4.10) we see that the e-invariant of $\delta(j_\#)$ is defined and given by

$$\mathbf{e}(\delta(j_\#)) = \mathbf{d}(\beta^{-1}\delta(j_\#)). \tag{4.11}$$

So if u is an element of $\tilde{\omega}^{-2}(S(\phi)^{\pi^*(\nu-\phi)})$, we have

$$\begin{aligned} \mathbf{e}(j^*\delta(u)) &= \mathbf{e}(u \cdot \delta(j_\#)) \\ &= \mathbf{d}(u) \cdot \mathbf{e}(\delta(j_\#)) \end{aligned} \tag{4.12}$$

(where the dot is the composition product). But, since $\pi^* : \tilde{K}^{-2}(P(\mathbf{C}^m)^{\nu-\phi}) \to \tilde{K}^{-2}(S(\phi)^{\pi^*(\nu-\phi)})$ is surjective, we have $\mathbf{d}(u) = \pi^*(v)$ for some v, or equivalently

$$\mathbf{d}(u) = v \cdot \mathbf{d}(\pi_\#), \tag{4.13}$$

where $\pi_\# \in \omega^0\{S(\phi)^{\pi^*(\nu-\phi)}; P(\mathbf{C}^m)^{\nu-\phi}\}$ is the class given by the projection π. Assembling (4.11), (4.12) and (4.13), we obtain

$$\mathbf{e}(j^*\delta(u)) = v \cdot \mathbf{d}(\pi_\# \cdot \beta^{-1}\delta(j_\#)). \tag{4.14}$$

In the diagram (4.10) the element $\pi_\# \cdot \beta^{-1}\delta(j_\#)$ is given by the downward zigzag $\pi\beta^{-1}\delta$. The upward zigzag yields the alternative description $(\rho\gamma^{-1}\tau)(j_\#)$ and the formula $\mathbf{e}(j^*\delta(u)) = v \cdot \mathbf{d}((\rho\gamma^{-1}\tau)(j_\#)) \in (\tilde{K}\mathbf{Q}/\mathbf{Z})^{-2}(P(\mathbf{C}^n)^\nu)$ or

$$\mathbf{e}(j^*\delta(u)) = [v \cdot \mathbf{d}((\gamma^{-1}\tau)(j_\#))] \in (\tilde{K}^{-2}(P(\mathbf{C}^n)^\nu) \otimes \mathbf{Q})/\tilde{K}^{-2}(P(\mathbf{C}^n)^\nu). \tag{4.15}$$

This reduces the determination of the e-invariant (4.7) to a computation in rational cohomology. In the Gysin sequence for $S(F)$, the group $K_{\mathbf{T}}^{-2}\{N^+; F^+\}$ maps onto $K_{\mathbf{T}}^{-1}\{N^+; S(F)_+\}$. We have, therefore, to compute the composition:

$$K_{\mathbf{T}}^{-2}\{N^+; F^+\} \otimes \mathbf{Q} \longrightarrow \tilde{K}^{-2}(P(\mathbf{C}^m)^{\nu-\phi}) \otimes \mathbf{Q} \longrightarrow \tilde{K}^{-2}(P(\mathbf{C}^n)^\nu) \otimes \mathbf{Q}, \tag{4.16}$$

where the first map is given by lifting to $S(mE)$ and the second is induced by $\gamma^{-1}\tau(j_\#)$. Applying the Chern character and using the Thom isomorphism, we can replace (4.16) by the sequence:

$$\begin{array}{ccccc} \mathfrak{C}_{\mathbf{T}}^{-2}\{N^+; F^+\} & \longrightarrow & \tilde{\mathfrak{C}}^{-2}(P(\mathbf{C}^m)^{\nu-\phi}) & \longrightarrow & \tilde{\mathfrak{C}}^{-2}(P(\mathbf{C}^n)^\nu) \\ \| & & \| & & \| \\ \mathbf{Q}[[t]] & & \mathbf{Q}[[t]]/(t^m) & & \mathbf{Q}[[t]]/(t^n). \end{array} \tag{4.17}$$

Let $e_F \cdot t^{\dim F} \in \mathfrak{C}_{\mathbf{T}}^0(*) = \mathbf{Q}[[t]]$ be the element corresponding under periodicity to the cohomology Euler class of F in $H_{\mathbf{T}}^{2\dim F}(*)$. Then the second map in (4.17), induced by $\gamma^{-1}\tau(j_\#)$, is given by:

$$t^i \mapsto \begin{cases} 0, & \text{if } i < \dim F; \\ e_F^{-1} t^{i-\dim F}, & \text{if } i \geq \dim F. \end{cases} \tag{4.18}$$

This follows because γ in (4.10) is given by multiplication by the Euler class of ϕ:

$$\prod_i \mathrm{Hom}\,(\tilde{H}^i(P(\mathbf{C}^m)^{\nu-\phi};\mathbf{Q}),\tilde{H}^i(P(\mathbf{C}^n)^{\nu};\mathbf{Q})) \longrightarrow$$
$$\prod_i \mathrm{Hom}\,(\tilde{H}^i(P(\mathbf{C}^m)^{\nu};\mathbf{Q}),\tilde{H}^i(P(\mathbf{C}^n)^{\nu};\mathbf{Q})).$$

On the other hand, $\mathfrak{C}_{\mathbf{T}}^{-2}\{N^+;F^+\} \to \mathfrak{C}_{\mathbf{T}}^{-1}\{N^+;S(F)_+\} \to \mathfrak{C}_{\mathbf{T}}^{-1}\{N^+;(\mathbf{E}\mathfrak{F})_+\}$: $\mathbf{Q}[[t]] \twoheadrightarrow \mathbf{Q}[[t]]/(e_F\cdot t^{\dim F}) \rightarrowtail \mathbf{Q}((t))/\mathbf{Q}[[t]]$ is given by dividing by $e_F\cdot t^{\dim F}$. So, finally, we can read off the formula (4.5) from the expression (2.8) for the Chern character. ∎

Still following Knapp's programme for free $\mathbf{T}$-actions we consider the situation when the $\mathbf{T}$-action on M extends to an action of the torus $\mathbf{T}\times\mathbf{T}$.

4.19 Proposition. *Suppose that the action of the circle $\mathbf{T}$, included as $\mathbf{T}\times 1$ in the torus $\mathbf{T}\times\mathbf{T}$, extends to an action of $\mathbf{T}\times\mathbf{T}$ on M with finite isotropy groups. Suppose further that N admits a $\mathbf{T}\times\mathbf{T}$-structure for which the trivialization of $\hat{\tau}$ is equivariant. Then the K-theory invariant f_M, and so also the complex* e*-invariant, of M vanish.*

Proof. Every finite subgroup of $\mathbf{T}\times\mathbf{T}$ is contained in a subgroup of the form $\mathbf{T}(k)\times\mathbf{T}(k)$. It follows from the general theory of classifying spaces, as in [14: I.6], that there exists a $\mathbf{T}\times\mathbf{T}$-map

$$M \longrightarrow S(F_1)\times S(F_2)$$

for some complex $\mathbf{T}\times\mathbf{T}$-modules F_1 and F_2 with $F_1^{\mathbf{T}\times 1}=0$, $F_1^{1\times\mathbf{T}}=F_1$, $F_2^{1\times\mathbf{T}}=0$, $F_2^{\mathbf{T}\times 1}=F_2$. (The argument is not difficult. A tubular neighbourhood of an orbit in M can be retracted onto the orbit and so mapped to $\mathbf{T}/\mathbf{T}(k)\times\mathbf{T}/\mathbf{T}(k)$ for some $k\geq 1$. Choose a finite equivariant partition of unity subordinate to a covering of M by such tubular neighbourhoods. Then the required map can be constructed, one factor at a time, by forming joins: $S(E^{k_1})*\ldots*S(E^{k_n})=S(E^{k_1}\oplus\ldots\oplus E^{k_n})$.)

Now the $\mathbf{T}\times\mathbf{T}$-equivariant Pontrjagin-Thom construction produces a class in $K_{\mathbf{T}\times\mathbf{T}}^{-1}\{N^+;(S(F_1)\times S(F_2))_+\}$ lifting $f_M\in K_{\mathbf{T}}^{-1}\{N^+;(\mathbf{E}\mathfrak{F})_+\}$. The proof will be completed by showing that this group, isomorphic by duality and Bott periodicity to $K_{\mathbf{T}\times\mathbf{T}}^{-1}(S(F_1)\times S(F_2))$, is zero.

The representation ring $R(\mathbf{T}\times\mathbf{T})=K_{\mathbf{T}\times\mathbf{T}}^{0}(*)$ is the ring of Laurent polynomials in two generators z_1,z_2 lifted from $z\in R(\mathbf{T})$ by the projections onto the two factors. Write $p_i(z_i)$ for the K-theory Euler class $\sum(-1)^j[\Lambda^j F_i]$ of F_i. Then the Gysin sequence for $S(F_1)$ yields:

$$K_{\mathbf{T}\times\mathbf{T}}^{-1}(S(F_1))=0$$
$$K_{\mathbf{T}\times\mathbf{T}}^{0}(S(F_1))=R(\mathbf{T}\times\mathbf{T})/(p_1(z_1)). \tag{4.20}$$

Dealing similarly with $S(F_2)$, and using periodicity, we obtain an exact sequence:

$$K^{-1}_{\mathbf{T}\times\mathbf{T}}(S(F_1)) \to K^{-1}_{\mathbf{T}\times\mathbf{T}}(S(F_1)\times S(F_2)) \to K^{0}_{\mathbf{T}\times\mathbf{T}}(S(F_1)) \to K^{0}_{\mathbf{T}\times\mathbf{T}}(S(F_1)),$$

in which the last map is multiplication by $p_2(z_2)$. The vanishing of the group $K^{-1}_{\mathbf{T}\times\mathbf{T}}(S(F_1)\times S(F_2))$ then follows from (4.20) (because there is unique factorization in $R(\mathbf{T}\times\mathbf{T})$). ∎

Manifolds M of the type envisaged in 4.19 can be constructed as homogeneous spaces G/Γ where G is a Lie group containing a compact subgroup of rank at least 2 and Γ is a discrete subgroup. Embed $\mathbf{T}\times\mathbf{T}$ in G and let it act on M by left translation. A framing is provided by (3.3). (These examples have been extensively studied in [5], [19], [15] and elsewhere.)

4.21 Remark. The description 4.6 of the e-invariant in terms of the K-theory class f_M extends, with merely notational changes, to more general framings of the type: $N_0 \xrightarrow{\cong} \hat{\tau}\oplus N_1$ where N_0 and N_1 are complex $\mathbf{T}$-modules.

If the fixed subspace $N^{\mathbf{T}}$ of the module N is trivial, we set $i^*(x_M)=\sum a_n\sigma_n \in \omega_1^{\mathbf{T}}(E\mathfrak{F})$ as in 2.20. The stable homotopy invariants (a_n) determine f_M by 2.21 and hence the e-invariant of M, 4.3. For a 3-manifold, $r=1$, we obtain the following attractive formulae.

4.22 Corollary. *Suppose that $N=E$. Then*

$$f_M(z)=\sum a_n((1-z)(1-z^n))^{-1}-(\tfrac{1}{2}\sum a_n)(1-z)^{-1}\in S^{-1}R/R$$

and the complex e-invariant of M is $(\sum n\,a_n)/12 \in \mathbf{Q}/\mathbf{Z}$.

Proof. Use 2.23 and (2.25). ∎

4.23 Corollary. *Suppose that $N=E^k$, $k\geq 1$. Then the complex e-invariant of M is*

$$\left(\sum n\,a_n\right)/(12k)-\sum_{l\mid k}\left(\sum_{(n,k)=l} a_n s(n/l,k/l)\right)\in \mathbf{Q}/\mathbf{Z},$$

where $s(n/l,k/l)$ is the Dedekind sum.

Proof. The Dedekind sum $s(n,k)$ is defined for coprime n and k by

$$s(n,k)=\frac{1}{4k}\sum_{j=1}^{k-1}\cot((\pi j)/k)\cdot\cot((\pi jn)/k)=-\frac{1}{4k}\sum_{\substack{\zeta^k=1\\ \zeta\neq 1}}\frac{\zeta+1}{\zeta-1}\cdot\frac{\zeta^n+1}{\zeta^n-1}.$$

From the identity

$$\frac{\zeta+1}{\zeta-1}\cdot\frac{\zeta^n+1}{\zeta^n-1}=1+\frac{2\zeta}{(\zeta-1)(\zeta^n-1)}+\frac{2\zeta^{-1}}{(\zeta^{-1}-1)(\zeta^{-n}-1)},$$

one deduces the alternative expression:

$$-s(n,k) = (k-1)/4k + \frac{1}{k} \sum_{\substack{\zeta^k=1 \\ \zeta\neq 1}} \frac{1}{\zeta^n - 1} \cdot \frac{\zeta}{\zeta - 1}.$$

The formula 4.23 comes from (2.24) and (2.25). ∎

4.24 Remark. If, instead of (2.25), one uses (2.26) in the proof of 4.23, one obtains a different formula involving Dedekind sums. It is related to the first by Rademacher reciprocity. See, for example, [16], where there are also proofs of the facts that, for coprime n and k, $12k\, s(n,k)$ is an integer and its reduction $(\mathrm{mod}\ k)$ is equal to $n + n^{-1}$. The congruences 2.16 confirm that 12e, as given by 4.23, is zero.

In the next section we shall see that the invariants (a_n) are Euler characteristics and so often readily computable.

4.25 Remark. For simplicity, attention has been restricted to the complex e-invariant. One can also work in KO-theory to treat the real e-invariant. In many cases, depending on the dimension of N and whether or not $det(N)$ is a square, the real theory runs parallel to the complex. For example, if $N = E$ as in 4.22 one obtains the real e-invariant $(\sum na_n)/24$ (mod $\mathbf{Z}$). In other cases the KO-invariant includes a subtler mod 2 component. It is planned to treat this elsewhere.

5 The Euler characteristic

Let M now be an arbitrary closed manifold with fixed-point-free circle action. We consider the stable cohomotopy Euler class $\gamma(\hat{\tau}) \in \tilde{\omega}^0_{\mathbf{T}}(M^{-\hat{\tau}})$ of the normal bundle to the orbits. (For the definition and elementary properties of the Euler class, here denoted by γ, see, for example, [10] or [11].) The image of $\gamma(\hat{\tau})$ under duality and the classifying map $M \to \mathbf{E}\mathfrak{F}$:

$$\tilde{\omega}^0_{\mathbf{T}}(M^{-\hat{\tau}}) \cong \omega_1^{\mathbf{T}}(M) \longrightarrow \omega_1^{\mathbf{T}}(\mathbf{E}\mathfrak{F}) = \bigoplus \mathbf{Z}\sigma_n \tag{5.1}$$

is a basic invariant of the fixed-point-free $\mathbf{T}$-manifold M; we denote it by $\mathcal{E}(M)$ and can reasonably call it an *Euler characteristic*. (The true Euler characteristic, the image of $\gamma(\tau M)$ in $\omega_0^{\mathbf{T}}(*) = \mathbf{Z}$, is zero.)

5.2 Proposition. *The components of $\mathcal{E}(M)$ are given by*

$$\mathcal{E}(M) = \sum \chi_c(M_{(\mathbf{T}(n))}/\mathbf{T}).\sigma_n,$$

where χ_c denotes the Euler characteristic with compact supports and $M_{(\mathbf{T}(n))}$ the set of points in M with isotropy group exactly $\mathbf{T}(n)$.

For general results of this type see [14: IV]. The proof sketched below is similar to the proof in [6] by Becker and Schultz of an analogous result for a closely related Lefschetz index.

Outline proof of 5.2. When the circle action is free, the result follows readily from the classical relation between the Euler class of the tangent bundle of $M/\mathbf{T}$ and the Euler characteristic. We describe a reduction to this case.

In order to dissect M it is desirable to extend the invariant $\mathcal{E}$ to manifolds with boundary. Let W be a compact smooth manifold with fixed-point-free $\mathbf{T}$-action. We again write $\hat{\tau}$ for the normal bundle to the orbits of W. Since the orbits are tangent to the boundary ∂W, the outward normal gives a nowhere-zero section v of $\hat{\tau}|\partial W$, up to homotopy. The relative Euler class $\gamma(\hat{\tau}, v) \in \tilde{\omega}^0_{\mathbf{T}}((W, \partial W)^{-\hat{\tau}}) = \omega_1^{\mathbf{T}}(W)$ supplies, via the classifying map, an invariant $\mathcal{E}(W) \in \omega_1^{\mathbf{T}}(\mathbf{E}\mathfrak{F})$. It shares the following properties with the classical Euler characteristic.

5.3 Lemma. *(i) If ξ is a $\mathbf{T}$-vector bundle over M, $D\xi$ its disc bundle, then $\mathcal{E}(D\xi) = \mathcal{E}(M)$.*

(ii) If M has a decomposition $M = W_0 \cup W_1$ as a union of compact submanifolds W_0, W_1 intersecting in their common buondary $\partial W_0 = \partial W_1$ then

$$\mathcal{E}(W_0 \cup W_1) + \mathcal{E}(W_0 \cap W_1) = \mathcal{E}(W_0) + \mathcal{E}(W_1).$$

Proof. The arguments for the classical Euler characteristic carry through virtually unchanged. It is convenient to rephrase (5.1). Choose a map $f : M \to S(F)$, where F is a $\mathbf{T}$-module with $F^{\mathbf{T}} = 0$. Then (5.1) is given by the Umkehr map $f_!$ and duality for $S(F)$:

$$\tilde{\omega}^0_{\mathbf{T}}(M^{-\hat{\tau}}) = \tilde{\omega}^{-1}_{\mathbf{T}}(M^{-\tau M}) \xrightarrow{f_!} \tilde{\omega}^{-1}_{\mathbf{T}}(S(F)^{-\tau S(F)}) \cong \omega_1^{\mathbf{T}}(S(F)) \longrightarrow \omega_1^{\mathbf{T}}(\mathbf{E}\mathfrak{F}). \qquad (5.4)$$

The invariants in 5.3 all lift to $\tilde{\omega}^0_{\mathbf{T}}(M^{-\hat{\tau}})$, and the identities (i) and (ii) can be established at that level. Thus for (ii), one fattens $W_0 \cap W_1$ to a tubular neighbourhood and writes M as $W_0 \cup_\partial (Y \times [0,1]) \cup_\partial W_1$ with $\partial W_i = Y \times \{i\}$. The outward normals on the boundaries of W_0 and W_1 give a section v of $\hat{\tau}|Y \times \{0,1\}$. Then the relative Euler class $\gamma(\hat{\tau}, v)$ in $\tilde{\omega}^0((M, Y \times \{0,1\})^{-\hat{\tau}})$ splits into three components giving $\mathcal{E}(W_0)$, $-\mathcal{E}(Y)$, $\mathcal{E}(W_1)$ and restricts to $\gamma(\hat{\tau})$ in $\tilde{\omega}^0_{\mathbf{T}}(M^{-\hat{\tau}})$. ∎

The reduction in the proof of 5.2 now proceeds as follows. If M has more than a single orbit type, it admits a decomposition of the form 5.3(ii) with W_0 a tubular neighbourhood of a proper submanifold $M^{\mathbf{T}(m)}$, $m > 1$. The disc bundle W_0 can be handled using 5.3(i). Two copies of W_1 can be glued along the boundary to give a closed manifold with fewer orbit types than M. Repeated application of the additivity property, 5.3(ii), completes the proof. ∎

Suppose we now have, as in Section 4, a $\mathbf{T}$-equivariant trivialization: $M \times N \xrightarrow{\cong} \hat{\tau}$, where N is a complex $\mathbf{T}$-module with $N^{\mathbf{T}} = 0$. Consider the associated invariant $x_M \in \omega_{\mathbf{T}}^{-1}\{N^+; (\mathbf{E}\mathfrak{F})_+\}$ defined at the beginning of that section. The restriction map $i^* : \omega_{\mathbf{T}}^{-1}\{N^+; (\mathbf{E}\mathfrak{F})_+\} \to \omega_{\mathbf{T}}^{-1}\{0^+; (\mathbf{E}\mathfrak{F})_+\} = \omega_1^{\mathbf{T}}(\mathbf{E}\mathfrak{F})$, as in 2.20, can be thought of as multiplication by the Euler class $\gamma(N)$. It, therefore, follows immediately from the construction of x_M and $\mathcal{E}(M)$ that:

$$i^*(x_M) = \mathcal{E}(M). \tag{5.5}$$

Finally, let us apply the theory to the example (3.4). Write $\mathcal{E}(M) = \sum a_n \sigma_n$. By 5.2 we have

$$\sum a_n = \chi(M/\mathbf{T}) = 2(1-g), \tag{5.6}$$

where g equals the genus of the surface $M/\mathbf{T}$, and a_n, for $n > 1$, is the number $\chi(M_{(\mathbf{T}(n))}/\mathbf{T})$ of elliptic points of period n on $M/\mathbf{T}$.

The sequence (a_n) thus contains the information commonly called the signature of Γ. Note that the Hurewicz image of $\mathcal{E}(M)$ in integral homology $H_1^{\mathbf{T}}(E\mathfrak{F}) = \mathbf{Q}$ is, by 2.11,

$$\sum a_n/n = \chi(\Gamma), \tag{5.7}$$

the Euler characteristic of Γ. (See, for example, [8: IX].) From 4.22 we obtain the result of Seade-Steer [22] that: *the complex* e*-invariant of M (with the left-invariant framing) equals*

$$\sum n\, a_n/12 \pmod{\mathbf{Z}}. \tag{5.8}$$

5.9 **Remark.** It is not necessary to use all the machinery of Sections 2 and 4 to obtain (5.8). It is enough to describe the forgetful map: $\tilde{\omega}_{\mathbf{T}}^{-1}(E^+) \to \omega^{-3}(*) = \mathbf{Z}/24$. This may be obtained by a K-theory calculation using the J-map, $\tilde{J}$. Consider the commutative diagram:

$$\begin{array}{ccccc} K_{\mathbf{T}}^{-2}(*) & \xleftarrow{i^*} & \tilde{K}_{\mathbf{T}}^{-2}(E^+) & \longrightarrow & K^{-4}(*) \\ \downarrow{\scriptstyle \tilde{J}} & & \downarrow{\scriptstyle \tilde{J}} & & \downarrow{\scriptstyle \tilde{J}} \\ \omega_{\mathbf{T}}^{-1}(*) & \xleftarrow{i^*} & \tilde{\omega}_{\mathbf{T}}^{-1}(E^+) & \longrightarrow & \omega^{-3}(*). \end{array} \tag{5.10}$$

Put $\alpha_n = (1 + z + \ldots + z^{n-1})\lambda_{\mathbf{C}}.\lambda_E \in \tilde{K}_{\mathbf{T}}^{-2}(E^+)$. Then $i^*(\alpha_n) = (1-z^n)\lambda_{\mathbf{C}}$ and, hence, $i^*\tilde{J}(\alpha_n) = \sigma_n + \eta$. On the other hand, α_n maps to n times a generator of $K^{-4}(*)$ and so of $\omega^{-3}(*)$.

References

[1] J.F. Adams: On the Chern character and the structure of the unitary group. *Proc. Cambridge Philos. Soc.* 57 (1961), 189–199.

[2] M.F. Atiyah: Elliptic operators and compact groups. Lecture Notes in Math. 401, Springer, 1974.

[3] M.F. Atiyah and I.M. Singer: The index of elliptic operators. I. *Ann. of Math.* 87 (1968), 484–530.

[4] M.F. Atiyah and I.M. Singer: The index of elliptic operators. III. *Ann. of Math.* 87 (1968), 546–604.

[5] M.F. Atiyah and L. Smith: Compact Lie groups and the stable homotopy of spheres. *Topology* 13 (1974), 135–142.

[6] J.C. Becker and R.E. Schultz: Fixed point indices and left invariant framings. *Geometric Applications of Homotopy Theory I, pp. 1–31, Lecture Notes in Math. 657, Springer*, 1978.

[7] B. Blackadar: K-theory for Operator Algebras, Springer, 1986.

[8] K.S. Brown: Cohomology of Groups, Springer, 1982.

[9] M.C. Crabb: $\mathbf{Z}/2$–Homotopy Theory, Cambridge University Press, 1980.

[10] M.C. Crabb and K.Knapp: On the codegree of negative multiples of the Hopf bundle. *Proc. Roy. Soc. Edinburgh* 107A (1987), 87–107.

[11] M.C. Crabb and W.A. Sutherland: The space of sections of a sphere-bundle. I. *Proc. Edinburgh Math. Soc.* 29 (1986), 383–403.

[12] T. tom Dieck: Bordism of G-manifolds and integrality theorems. *Topology* 9 (1970), 345–358.

[13] T. tom Dieck: Transformation groups and representation theory. Lecture Notes in Math. 766, Springer, 1979.

[14] T. tom Dieck: Transformation groups, Walter de Gruyter, 1987.

[15] H.D. Fegan, B. Steer and L. Whiteway: On the spectral symmetry of the Dirac operator on compact quotients $\Gamma\backslash G$ of Lie groups G. Preprint.

[16] F. Hirzebruch and D. Zagier: The Atiyah-Singer Theorem and Elementary Number Theory, Publish or Perish, 1974.

[17] J.D.S. Jones: Cyclic homology and equivariant homology. *Invent. Math.* 87 (1987), 403–423.

[18] K. Knapp: Das Bild des Hurewicz-Homomorphismus $h\colon \pi_*^s(B\mathbf{Z}_p) \to K_1(B\mathbf{Z}_p)$. *Math. Ann.* 223 (1976), 119–138.

[19] K. Knapp: Rank and Adams filtration of a Lie group. *Topology* 17 (1978), 41–52.

[20] K. Knapp: Some applications of K-theory to framed bordism, Habilitationsschrift, Bonn, 1979.

[21] J. Seade and B. Steer: A note on the eta function for quotients of $PSL_2(\mathbf{R})$ by co-compact Fuchsian groups. *Topology* 26 (1987), 79–91.

[22] J. Seade and B. Steer: The framed cobordism class of quotients of 3-dimensional Lie groups by co-compact discrete subgroups. Preprint.

[23] G. Segal: K-homology and algebraic K-theory. *K-theory and Operator algebras, pp. 113–127, Lecture Notes in Math. 575, Springer*, 1977.

Bitangent spheres and critical points

Duan Hai-bao and Elmer Rees

1 Background

The following theorem (announced at Cortona by the second author) answers a question raised by J.C. Bruce, P.J. Giblin, and C.G. Gibson [1].

Theorem. *Let $M^k \subset \mathbb{R}^{n+1}$ be a compact, smooth submanifold without boundary. If $x \in M$ and $\mathbf{n}$ is a unit vector normal to M at x, then there is a sphere Σ^n tangent to M at x and at some other point y and $\mathbf{n}$ is normal to Σ at x.*

The sphere Σ^n must be allowed to have infinite radius (in other words, to be a hyperplane). This is even necessary when M is a hypersurface (that is, when $k = n$). For suppose that E^n is an ellipsoid with a unique major axis whose end points are x_1 and x_2. There is clearly a unique sphere Σ that is bitangent to E at x_1 or x_2. Consider a conformal transformation σ of $\mathbb{R}^{n+1}$ such that $M = \sigma E$ is compact and which sends Σ to a hyperplane. The manifold M now has the property that the only bitangent sphere to it at either x_1 or x_2 is the hyperplane $\sigma\Sigma$.

The proof of the theorem involves the study of the critical points of the function $f: M - \{x\} \to \mathbb{R}$, where $f(y)$ equals the (signed) curvature of the unique sphere passing through y and normal to $\mathbf{n}$ at x. It is a simple geometrical fact that if y is a critical point of f then the corresponding sphere is tangential to M at y. It therefore remains to show that f must have a critical point. Because the normal curvature of M at x is given by a quadratic expression, the function f extends to a function $\tilde{f} : \tilde{M} \to \mathbb{R}$ where $\tilde{M}$ is the (real) blow up of M at x. A study of the Lusternik-Schnirelmann category (for references see [4] and [5]) shows that it is not possible for all the critical points of $\tilde{f}$ to be on the $\mathbb{R}P^{k-1} \subset \tilde{M}$ that is the blow up of x. Hence the function f must have a critical point y. If $f(y) = 0$, then the corresponding bitangent sphere would be a hyperplane.

The proof of the theorem has now appeared in [2], and in the next section, we describe some results suggested by some of our earlier efforts to prove this theorem. Preliminary versions of these were also discussed at the conference, and full details will appear in a future paper.

2 Results

Whilst attempting to prove the above theorem, we studied functions on manifolds that had preassigned critical sets (the function $\tilde{f}$ only has real projective spaces as its critical sets on the blow up of x). This seems a natural problem in its own right. A few theorems of this kind have already been proved. If the function $f: M \to \mathbf{R}$ has a single maximum and a single minimum and M is compact without boundary then M is homeomorphic to a sphere (for references see [7, page 25]). When f has three critical points the situation has been studied by J. Eells and N.H. Kuiper [3]. These are the only results which we are aware of in the literature. We have obtained the following.

Let M^n be a compact connected smooth manifold without boundary and let $f: M \to \mathbf{R}$ be a smooth function whose critical set is $V_0 \sqcup V_1$, where V_0 and V_1 are both connected smooth submanifolds of M. (No assumption concerning the non-degeneracy of the function at the critical sets is made.) The manifold $V_0 = f^{-1}(0)$ is the set where f attains its minimum and $V_1 = f^{-1}(1)$ that where f attains its maximum. By considering the function $-f$ the roles of V_0 and V_1 can be interchanged.

2.1 If $\dim V_0 = \dim V_1 = 0$, then, as mentioned above, M is homeomorphic to a sphere.

2.2 If $\dim V_0 = 0$ and $\dim V_1 > 0$, then M is homotopy equivalent to one of the projective spaces (over $\mathbf{R}, \mathbf{C}, \mathbf{H}$ or $\mathbf{O}$), and V_1 is a codimension one projective space.

2.3 If V_0 is a homotopy $\mathbf{R}P^k$ and $\dim V_1 < k$, then the pair $M; V_1$ must be one of the following (# denotes connected sum):

$$\mathbf{R}P^{k+1} \# \mathbf{C}P^{m+1};\ \mathbf{C}P^m \quad \text{with } k = 2m+1,$$

$$\mathbf{R}P^{k+1} \# \mathbf{H}P^{m+1};\ \mathbf{H}P^m \quad \text{with } k = 4m+3,$$

$$\mathbf{R}P^{16} \# \mathbf{O}P^2;\ \mathbf{O}P^1 = S^8 \quad \text{with } k = 15,$$

$$\mathbf{R}P^{k+\ell+1};\ \mathbf{R}P^\ell \quad \text{with } \ell < k.$$

It is clear that all these cases can occur.

In the case where V_0 and V_1 are both $\mathbf{R}P^k$, then a number of possibilities can occur for M (as in the special case $k = 0$ already). For example, M could be $S^n \times \mathbf{R}P^k$ for any $n \geq 1$, or it could be $\mathbf{R}P^{k+1} \# \mathbf{R}P^{k+1}$. A more interesting case is when $k = 2$ and one can have $M = S^4$.

Regard S^4 as the space of 3×3 real symmetric matrices A with $\operatorname{tr} A = 0$ and $\operatorname{tr} A^2 = 1$ (those with $\operatorname{tr} A = 0$ form a 5-dimensional real vector space, and $\operatorname{tr} A^2 = 1$ is a positive definite quadratic form). Using Lagrange multipliers one can easily check that the functional $\operatorname{tr} A^3$ has critical set $\mathbf{R}P^2 \sqcup \mathbf{R}P^2$ (two copies of the Veronese surface). For similar examples see [6].

References

[1] J.C. Bruce, P.J. Giblin, and C.G. Gibson: Symmetry sets. *Proc. Roy. Soc. Edinburgh* 101A (1985), 163–186.

[2] Duan H. and E. Rees: The existence of bitangent spheres. *Proc. Roy. Soc. Edinburgh* 111A (1989) 85–87.

[3] J. Eells and N.H. Kuiper: Manifolds which are like projective planes. *Inst. Hautes Études Sci. Publ. Math.* 14 (1962), 5–46.

[4] A.T. Fomenko: Differential Geometry and Topology. Consultants Bureau, New York, 1987.

[5] I.M. James: On category, in the sense of Lusternik-Schnirelmann. *Topology* 17 (1978), 331–348.

[6] W.S. Massey: Imbeddings of projective planes and related manifolds into spheres. *Indiana Univ. Math. J.* 23 (1973/74), 791–812.

[7] J. Milnor: Morse Theory. Ann. of Math. Stud. 51, Princeton University Press, 1963.

The enveloping ring of a Π-algebra

W. G. Dwyer and D. M. Kan*

1 Introduction

1.1 Summary. In this paper we start an investigation of Π-algebras, i.e. (≥ 1)-graded groups with an action of the primary homotopy operations.

Given a Π-algebra X, we construct an *enveloping ring* EX of X, which generalizes the integral group ring of a group, as well as the enveloping algebra of a connected graded rational algebra. We also obtain a *Milnor-Moore theorem* which gives a sufficient condition (*E-flatness*) on the homotopy Π-algebra π_*L of a pointed connected CW-complex L, in order that the *integral homology ring* $H_*\Omega L$ of the loop space ΩL of L be canonically isomorphic to the *enveloping ring* $E\pi_*L$ of π_*L.

Just as one uses the group ring of a group and the enveloping algebra of a Lie algebra to define their homology and cohomology, we will in [4] use the enveloping ring construction of this paper to develop the homology and cohomology of Π-algebras.

1.2 A more detailed outline. (i) *Π-algebras.* After fixing some notation and terminology (in 1.3), we recall (in Section 2) the definition of a Π-algebra and note that "aspherical" Π-algebras are essentially groups and that "simply connected rational" Π-algebras can (after lowering the degrees by 1) be considered as connected graded rational Lie algebras.

(ii) *The Milnor-Moore map.* In preparation for the construction of the enveloping ring of a Π-algebra we next (in Section 3) consider, for a pointed connected CW-complex L, its Milnor-Moore map $\mu: \pi_*L \to H_{*-1}\Omega L$, which we define as the composition of the boundary isomorphism $\pi_*L \approx \pi_{*-1}\Omega L$ with the Hurewicz map $\pi_*\Omega L \to H_*\Omega L$. This Milnor-Moore map lowers dimensions by 1 and is what we will call a **ΠR**-*map*, i.e. it is, in a certain precise manner, compatible with the Π-algebra structure of π_*L and the ring structure of $H_*\Omega L$. If L is aspherical (i.e. L has the same homotopy type as $K(\pi_1 L, 1)$), then the Milnor-Moore map becomes just the usual inclusion $\pi_1 L \to H_*\Omega L = \mathbb{Z}\pi_1 L$ of a group in its integral group ring. If L

*This research was in part supported by the National Science Foundation

is simply connected and rational (i.e. its higher homotopy groups are rational vector spaces), then $H_*\Omega L$ is the enveloping algebra of π_*L, considered (see (i)) as a connected graded rational Lie algebra, and the Milnor-Moore map becomes the inclusion of one in the other which these authors discussed in [7].

(iii) *The enveloping ring.* In Section 4, we define the enveloping ring EX of a $\mathbf{\Pi}$-algebra X, which (see (i)) reduces in the aspherical case to the integral group ring of X_1 (the group in degree 1) and in the simply connected rational case to the enveloping algebra of X (considered as a connected graded rational Lie algebra). Like group rings and enveloping algebras, enveloping rings can be characterized by

(iv) *A universal property.* The enveloping ring EX of a $\mathbf{\Pi}$-algebra X comes (also in Section 4) with a natural $\mathbf{\Pi R}$-map $e: X \to EX$, which is initial among the $\mathbf{\Pi R}$-maps starting at X. Hence the (see (ii)) Milnor-Moore map $\mu: \pi_*L \to H_{*-1}\Omega L$ factors through a *canonical ring homomorphism* $\mu': E\pi_*L \to H_*\Omega L$. If L is aspherical or simply connected and rational, then this canonical ring homomorphism is actually an isomorphism.

(v) *E-flat $\mathbf{\Pi}$-algebras and a Milnor-Moore theorem.* In Section 5 we prove a Milnor-Moore theorem, which gives a sufficient condition on the $\mathbf{\Pi}$-algebra π_*L, in order that the canonical ring homomorphism $\mu': E\pi_*L \to H_*\Omega L$ be an isomorphism. This sufficient condition is that the $\mathbf{\Pi}$-algebra π_*L be E-flat, i.e. that, for $p > 0$, the left derived functors $\mathbf{L}_pE$ of the enveloping ring functor E vanish on π_*L. The Milnor-Moore theorem then follows from the fact that the canonical map $\mu': E\pi_*L \to H_*\Omega L$ is the edge homomorphism in a first quadrant *Milnor-Moore spectral sequence*, which converges to $H_*\Omega L$ and of which the E^2-term consists of the $(\mathbf{L}_pE)\pi_*L$ $(p \geq 0)$. If L is aspherical, simply connected and rational, or a wedge of spheres, then π_*L is E-flat and we end with observing that other such spaces can be constructed from these by means of a lemma which states that, for a short exact sequence $* \to U \to W \to V \to *$ of $\mathbf{\Pi}$-algebras, W is E-flat whenever U and V are E-flat and $\mathrm{Tor}_1^{\mathbf{Z}}(EU, EV) = 0$. This lemma we deduce from

(vi) *A first quadrant Eilenberg-Moore spectral sequence for enveloping rings.* In the last section (Section 6) we show that a short exact sequence $* \to U \to W \to V \to *$ of $\mathbf{\Pi}$-algebras gives rise to a first quadrant Eilenberg-Moore spectral sequence which converges to $(\mathbf{L}_*E)V$ and for which

$$E^2_{p,q} = \mathrm{Tor}_p^{(\mathbf{L}_*E)U}(\mathbf{Z}, (\mathbf{L}_*E)W)_q$$

and we then deduce from this the above mentioned lemma.

1.3 Notation, terminology, etc. (i) *Spaces.* Whenever we consider the homotopy $\mathbf{\Pi}$-algebra π_*L of a space L, such a space L is always assumed to be an object of the category $\mathbf{CW}_*$ of pointed connected CW-complexes. Of course, its loop space ΩL need not be connected and hence need not be in this category.

(ii) *Rings.* All rings in this paper will be (≥ 0)-graded, augmented over $\mathbb{Z}$ and associative (though not necessarily commutative). We denote the category of such rings by **R**.

(iii) *Whitehead products.* For $L \in \mathbf{CW}_*$ and $a \in \pi_{p+1}L$, $b \in \pi_{q+1}L$ $(p, q \geq 1)$, we will denote by $[a, b]$ the Whitehead product which differs by $(-1)^p$ from the one of [12, Ch. X]. As a result,

$$[a, b] + (-1)^{pq}[b, a] = 0$$

and, if $c \in \pi_{r+1}L$ $(r \geq 1)$, then the Jacobi identity becomes

$$(-1)^{pr}[a, [b, c]] + (-1)^{pq}[b, [c, a]] + (-1)^{qr}[c, [a, b]] = 0.$$

(iv) *The Hopf invariant.* [12, Ch. X]. Let $i_p \in \pi_p S^p$ denote the obvious generator and let $\mu: \pi_* S^p \to H_{*-1}\Omega S^p$ be as in 1.2(ii). Then $H_*\Omega S^p$ $(p > 1)$ is the free ring (see (ii)) on a single generator μi_p in degree $p - 1$. If $\alpha \in \pi_r S^p$ $(r > p > 1)$, then $\mu\alpha = 0$, unless p is even and $r = 2p - 1$ (because $\pi_r S^p$ $(r > p > 1)$ is finite, unless p is even and $r = 2p - 1$). If p is even and $\alpha \in \pi_{2p-1}S^p$, then $\mu\alpha = H(\alpha)(\mu i_p)^2$ for some integer $H(\alpha)$. This integer $H(\alpha)$ is called the Hopf invariant of α.

2 Π-algebras

In this section we briefly review the notion of a **Π**-algebra. Its definition in 2.2 (a contravariant functor from the category **Π** of homotopy operations to the category of pointed sets, which sends coproducts to products) is elegant and sometimes useful in showing that certain constructions indeed yield **Π**-algebras. However, most of the time it is convenient to consider a **Π**-algebra as (2.3) a (≥ 1)-graded group with some additional structure.

We start with a description of a "category **Π** of (primary) homotopy operations" which is slightly different from, although equivalent to, the one of [11].

2.1 The category Π of homotopy operations. This will be the category which has as objects the spaces $M \in \mathbf{CW}_*$ (1.3(i)) with the homotopy type of a finite wedge of spheres (of dimensions ≥ 1), and which has as maps the (pointed) homotopy classes of maps between them.

Clearly the category **Π** is *pointed* and has *finite coproducts* (i.e. finite wedges).

2.2 Definition of Π-algebras. Let $\mathbf{Set}_*$ denote the category of pointed sets. A **Π**-*algebra* then is a contravariant functor $X: \mathbf{\Pi} \to \mathbf{Set}_*$ which sends coproducts to products and a map of **Π**-algebras is a natural transformation between two such functors. The resulting category of **Π**-algebras will be denoted by **Π-al**.

This definition immediately implies that, for every object $X \in \mathbf{\Pi\text{-}al}$,

(i) $X* = *$, *where* $*$ *denotes the point in both categories* $\mathbf{\Pi}$ *and* $\mathbf{Set}_*$, *and*

(ii) *the values of* X *on the objects of* $\mathbf{\Pi}$ *are determined by its value on the spheres* S^p $(p \geq 1)$.

In view of Hilton's analysis of the homotopy groups of wedges of spheres [12, Ch. XI] one therefore has, if one writes X_p for the value of X on S^p, the following

2.3 Convenient description of Π-algebras. A $\mathbf{\Pi}$-algebra X can be considered as a (≥ 1)-graded group $\{X_p\}_{p=1}^{\infty}$, with X_p abelian for $p > 1$, together with three kinds of operations

(i) a *Whitehead product* homomorphism $[\text{-},\text{-}]: X_p \otimes X_q \to X_{p+q-1}$ for all $p, q \geq 2$, and

(ii) a *composition* function $(\text{-}\circ\alpha): X_p \to X_r$ for every element $\alpha \in \pi_r S^p$ with $r > p > 1$ (which need not be a homomorphism, but which is right additive, i.e. $(x \circ \alpha_1) + (x \circ \alpha_2) = x \circ (\alpha_1 + \alpha_2)$ for all $x \in X_p$ and $\alpha_1, \alpha_2 \in \pi_r S^p$), which satisfy all the identities that hold for the Whitehead product and composition operations on the higher homotopy groups of pointed CW-complexes, and

(iii) a *left action* of X_1 on the X_p $(p > 1)$ which commutes with the Whitehead product and composition operations.

Thus a $\mathbf{\Pi}$-*algebra* X *is completely determined by its "universal covering"* $\tilde{X}$ (i.e. the sub $\mathbf{\Pi}$-algebra $\tilde{X} \subset X$ consisting of all the elements of X of degree > 1) *and the left action of the group* X_1 *on this* $\mathbf{\Pi}$-*algebra* $\tilde{X}$. We will denote by $\tau_x y$ the result of this left action by an element $x \in X_1$ on an element $y \in \tilde{X}$.

An obvious example of a $\mathbf{\Pi}$-algebra is provided by

2.4 The homotopy Π-algebra of a pointed connected CW-complex. Let $L \in \mathbf{CW}_*$ (1.3(i)). Then the functor $\mathbf{\Pi} \to \mathbf{Set}_*$ which sends an object $M \in \mathbf{\Pi}$ to the set of homotopy classes of maps $M \to L \in \mathbf{CW}_*$, is clearly a $\mathbf{\Pi}$-algebra in the sense of 2.2. In view of 2.3 this $\mathbf{\Pi}$-algebra can also be considered as the graded group $\{\pi_p L\}$ together with the usual Whitehead product and composition operations and fundamental group actions, and we will therefore denote this $\mathbf{\Pi}$-algebra by $\pi_* L$.

We end with a few more

2.5 Examples. (i) *Aspherical* $\mathbf{\Pi}$-*algebras.* These are $\mathbf{\Pi}$-algebras X such that (see 2.3) $\tilde{X} = *$. Clearly such a $\mathbf{\Pi}$-algebra is completely determined by the group X_1 and, conversely, every group determines such a $\mathbf{\Pi}$-algebra. *Aspherical* $\mathbf{\Pi}$-*algebras thus are just groups.*

(ii) *Simply-connected* $\mathbf{\Pi}$-*algebras.* These are the $\mathbf{\Pi}$-algebras X such that $X_1 = 1$. Such a $\mathbf{\Pi}$-algebra has an *"underlying" connected graded Lie algebra*, obtained by lowering the degrees by 1 and taking the Whitehead products as Lie products (see 1.3(iii)).

(iii) *Simply connected rational* **Π**-*algebras.* These are the simply connected **Π**-algebras X such that each X_p $(p > 1)$ is a rational vector space. If $x \in X_p$ and $\alpha \in \pi_r S^p$ $(r > p > 1)$, then (1.3(iv)) $x \circ \alpha = 0$ unless p is even and $r = 2p - 1$, in which case $2(x \circ \alpha) = H(\alpha)[x, x]$. The **Π**-algebra structure of X thus can be recovered from its (see(ii)) underlying Lie algebra, and it is not difficult to see that this correspondence yields an *equivalence between the category of simply connected rational* **Π**-*algebras and the category of connected rational Lie algebras.*

(iv) *Free* **Π**-*algebras.* These are the homotopy **Π**-algebras of wedges of spheres. If $M = \bigvee_{j \in J} S^{p_j}$ with all $p_j \geq 2$, then $\pi_* M$ is the free **Π**-algebra with, for each $j \in J$, a generator $i_{p_j} \in \pi_{p_j} S^{p_j} \subset \pi_{p_j} M$ in degree p_j, and Hilton's calculations [12, Ch. XI] show how each element of $\pi_* M$ can be obtained from these generators by taking an iterated Whitehead product, possibly followed by a composition operation. More generally, if $L = M \vee N$, where M is as before and N is a wedge of circles, then $\pi_* L$ is the free **Π**-algebra with a generator in degree 1 for each circle in N and, for each $j \in J$ a generator i_{p_j} in degree p_j. To describe the elements of $\pi_* L$ one then notes that the universal covering of $\pi_* L$ is just the homotopy **Π**-algebra $\pi_* \tilde{L}$ of the universal covering $\tilde{L}$ of L, which is the free **Π**-algebra with (see 2.3) a generator $\tau_x i_{p_j}$ of degree p_j for every $x \in \pi_1 N$ and $j \in J$, and that $\pi_1 N$ acts from the left on this **Π**-algebra by permuting the generators in the obvious manner.

(v) *Sub* **Π**-*algebras of free* **Π**-*algebras.* Clearly a sub **Π**-algebra of a free **Π**-algebra (iv) need not be free. However, if U and V are free **Π**- algebras and X is the kernel of the projection $U \sqcup V \to V$, then X is also free. More precisely, if $\{a_i\}_{i \in I}$ and $\{b_j\}_{j \in J}$ are sets of (free) generators for U and V respectively, then a lengthy but essentially straightforward calculation (using the corollary to [6, Th. 3]) shows that X is freely generated by the elements

$$v a_i v^{-1} \qquad \text{with } v \in V_1 \text{ and } a_i \in U_1$$

and the elements

$$[[\cdots[\tau_{v_0} a_{i_0}, \tau_{v_1} b_{j_1}], \cdots], \tau_{v_n} b_{j_n}] \quad \text{and}$$
$$[[\cdots[(\tau_{v a_i v^{-1} v_0} b_{j_0} - \tau_{v_0} b_{j_0}, \tau_{v_1} b_{j_1}], \cdots], \tau_{v_n} b_{j_n}]$$

with $n \geq 0$, $a_i \in U_1$, $a_{i_0} \in \tilde{U}$, $v, v_0, \ldots, v_n \in V_1$ and $b_{j_0}, \ldots, b_{j_n} \in \tilde{V}$.

It is also possible to give a direct geometric proof that X is free as follows. Let A and B be wedges of spheres (of dimensions ≥ 1) such that $\pi_* A = U$ and $\pi_* B = V$. Then the homotopy fibre of the collapse map $A \vee B \to B$ is the space $(A \times \Omega B)/(* \times \Omega B)$ (cf. [5, proof of 1.1]), which is naturally homotopy equivalent to $A \vee (A \wedge \Omega B)$ (see [12, p.519] for the case that A is simply connected; the general case then can be obtained by using a Barratt-Puppe sequence [12, p.130]

$$A' \times * \to (A' \times \Omega B)/(* \times \Omega B) \to A' \wedge \Omega B \to A \to \cdots$$

in which A' is such that $SA' = A$, and observing that the connecting map $A' \wedge \Omega B \to A$ is null homotopic). Moreover $A \wedge \Omega B$ is a wedge of $A \wedge \Omega \tilde{B}$ and copies of the space $(A \times \Omega \tilde{B})/(* \times \Omega \tilde{B})$ which (see above) is naturally homotopy equivalent to $A \vee (A \wedge \Omega \tilde{B})$ and the desired result now follows from the fact that [12, p.335] $A \wedge \Omega \tilde{B}$ splits, up to homotopy, as a wedge of spheres.

3 The Milnor-Moore map

In preparation for the definition of the enveloping ring of a Π-algebra (in Section 4), we discuss here

3.1 The Milnor-Moore map $\mu: \pi_* L \to H_{*-1}\Omega L$. Let $L \in \mathbf{CW}_*$ (1.3(i)). Then its *Milnor-Moore map* $\mu: \pi_* L \to H_{*-1}\Omega L$ will be the function obtained by composing the boundary isomorphism $\pi_* L \approx \pi_{*-1}\Omega L$ with the Hurewicz map $\pi_* \Omega L \to H_* \Omega L$.

While the Hurewicz map $\pi_* L \to H_* L$ kills all elements which are (see 2.3) in the image of the Whitehead products and the composition operations and all elements of the form $\tau_x y - y$ with $x \in \pi_1 L$ and $y \in \pi_p L$ ($p > 1$), this need not be the case for the Milnor-Moore map. To describe the behaviour of the Milnor-Moore map with respect to the Π-algebra structure of $\pi_* L$ and the ring structure of the integral homology ring $H_* \Omega L$ of the loop space of L, it is convenient to use the notion of

3.2 ΠR-maps. Given $X \in \mathbf{\Pi}\text{-}\mathbf{al}$ and $R \in \mathbf{R}$ (1.3(ii)), a function $f: X \to R$ will be called a *ΠR-map* if it lowers the degrees by 1 and satisfies the identities (see 1.3 and 2.3)

$$
\begin{aligned}
f(x+y) &= (fx)+(fy) && x, y \in \tilde{X}; \\
f[x,y] &= (fx)(fy) - (-1)^{pq}(fy)(fx) && x \in \tilde{X}_{p+1},\ y \in \tilde{X}_{q+1}; \\
f(x \circ \alpha) &= H(\alpha)(fx)^2 && \alpha \in \pi_{2p-1}S^p,\ p \text{ even}, \\
&= 0 && \text{otherwise}; \\
f(xy) &= (fx)(fy) && x, y \in X_1; \\
f(\tau_x y) &= (fx)(fy)(fx^{-1}) && x \in X_1,\ y \in \tilde{X}.
\end{aligned}
$$

This definition readily [12, Ch. X] implies

3.3 Proposition. *Let $L \in \mathbf{CW}_*$. Then the Milnor-Moore map $\mu: \pi_* L \to H_{*-1}\Omega L$ is a ΠR-map.*

We end with a few

3.4 Examples. (i) L *is aspherical* (i.e. has the same homotopy type as $K(\pi_1 L, 1)$). Then $H_i \Omega L = 0$ for $i > 0$ and $H_0 \Omega L = \mathbf{Z}\pi_1 L$, the integral group ring of

$\pi_1 L$, and the Milnor-Moore map $\mu: \pi_1 L \to H_0 \Omega L = \mathbf{Z}\pi_1 L$ is just the canonical map which sends the elements of $\pi_1 L$ to the corresponding (additive) generators of $\mathbf{Z}\pi_1 L$.

(ii) *L is simply connected and rational* (i.e. its higher homotopy groups are rational vector spaces). Then (2.5(iii)) $\pi_* L$ is a simply connected rational **Π**-algebra and [7, App.] $H_* \Omega L$ is the enveloping algebra of the underlying connected graded rational Lie algebra of $\pi_* L$ under the Milnor-Moore map $\mu: \pi_* L \to H_{*-1} \Omega L$.

(iii) $L = \bigvee_{j \in J} S^{p_j}$ *with all* $p_j \geq 2$. The case that J contains only one element was already discussed in 1.3(iv). In general (see 2.5) $\pi_* L$ is a free **Π**-algebra on generators $i_{p_j} \in \pi_{p_j} S^{p_j} \subset \pi_{p_j} L$ (in degrees p_j) and [3] $H_* \Omega L$ is the free ring on generators μi_{p_j} (in degrees $p_j - 1$). It follows that *an element $x \in \pi_* L$ is of finite order if and only if $\mu x = 0$.*

4 The enveloping ring of a Π-algebra

We now define the enveloping ring of a **Π**-algebra, list some of its most immediate properties and discuss a few examples.

4.1 The enveloping ring of a Π-algebra. The *enveloping ring* of a **Π**-algebra X will be the ring (1.3(ii)) EX which has, for every integer $p \geq 1$ and element $x \in X_p$, a generator ex in degree $p-1$ with as relations the relations obtained from 3.2 by replacing everywhere f by e.

Clearly this definition is functorial, and *the natural function $e: X \to EX$, given by $x \to ex$ for all $x \in X$, is a* **ΠR**-*map* (3.2) with the

4.2 Universal property. *Let $X \in$ **Π-al** and $R \in$ **R** and let $f: X \to R$ be a* **ΠR**-*map. Then there is a unique map $f': EX \to R \in$ **R** such that $f'e = f: X \to R$.*

In view of 3.3 this implies

4.3 Corollary. *Given $L \in$ **CW**$_*$ (see 1.3(i)), there is a unique map $\mu': E\pi_* L \to H_{*-1} \Omega L \in$ **R** such that $\mu' e = \mu: \pi_* L \to H_{*-1} \Omega L$.*

4.4 Examples. (i) If $X \in$ **Π-al** is *aspherical* (2.5(i)), then $EX = \mathbf{Z}X_1$, the integral group ring of X_1, and the map $e: X_1 \to EX = \mathbf{Z}X_1$ is the usual map which sends the elements of X_1 to the corresponding (additive) generators of $\mathbf{Z}X_1$.

(ii) If $X \in$ **Π-al** is simply connected rational (2.5(iii)), then EX is the enveloping algebra of X and the map $e: X \to EX$ is the usual inclusion of one in the other.

(iii) If $X \in$ **Π-al** *is simply connected and free,* (i.e. $X = \pi_* \bigvee_{j \in J} S^{p_j}$ with all $p_j \geq 2$), then, in the notation of 2.5(iv), EX is the free ring on generators ei_{p_j} in degrees $p_j - 1$.

(iv) *If $L \in$ **CW**$_*$ is aspherical, simply connected and rational or a wedge of spheres of dimension at least 2, then,* in view of (i)–(iii) above and 3.4, *the canonical map*

$\mu': E\pi_*L \to H_*\Omega L \in \mathbf{R}$ *of 4.3 is an isomorphism.* We will see in Section 5 how one can construct many other such spaces.

Other immediate consequences of Definition 4.1 are

4.5 Proposition. *The functor E: $\mathbf{\Pi}$-al $\to R$ preserves direct limits.*

4.6 Proposition. *Let $X \in \mathbf{\Pi}$-al, and let $\tilde{X} \subset X$ be its universal covering (2.3). Then EX is the "semi tensor product" of $E\tilde{X}$ and $\mathbf{Z}X_1$, in the sense that*

(i) additively EX is naturally isomorphic to $E\tilde{X} \otimes \mathbf{Z}X_1$, and

(ii) the multiplication in EX is given by

$$(ey \otimes ex)(ey' \otimes ex') = (ey)(e\tau_x y') \otimes (ex)(ex')$$

for all $x, x' \in X_1$ and $y, y' \in \tilde{X}$.

4.7 Proposition. *A short exact sequence $* \to U \to W \to V \to *$ of $\mathbf{\Pi}$-algebras induces a natural isomorphism $EW \otimes_{EU} \mathbf{Z} \approx EV$.*

5 E-flat Π-algebras and a Milnor-Moore theorem

In 4.4 we noted that for certain spaces $L \in \mathbf{CW}_*$, the canonical ring homomorphism $\mu': E\pi_*L \to H_*\Omega L$ of 4.3 actually is an isomorphism. We now obtain a Milnor-Moore theorem which states that a sufficient condition for this to happen is, that the $\mathbf{\Pi}$-algebra π_*L be "E-flat".

We start with observing that the canonical map $\mu': E\pi_*L \to H_*\Omega L \in \mathbf{R}$ is not always an isomorphism, and that the most one can say about it in general is, that it is the edge homomorphism in

5.1 A Milnor-Moore spectral sequence. *Given $L \in \mathbf{CW}_*$, there is a natural first quadrant spectral sequence $\{E^r_{p,q}\}$, which converges to $H_*\Omega L$ and for which*

(i) $E^2_{0,*} = E\pi_*L$, and

(ii) $E^2_{p,*} = (\mathbf{L}_p E)\pi_*L$ for $p > 0$,

where (see 6.1) $\mathbf{L}_p E$ denotes the p-th left derived functor of the enveloping ring functor E: $\mathbf{\Pi}$-al $\to R$.

Moreover, the edge homomorphism

$$E\pi_*L = E^2_{0,*} \to E^\infty_{0,*} \to H_*\Omega L$$

in this spectral sequence coincides with the canonical map $\mu': E\pi_*L \to H_*\Omega L$ of 4.3.

Proof. The construction of this spectral sequence and the verification of its properties are essentially the same as for the Hurewicz spectral sequence of [1], replacing everywhere $\mathbf{V}_\bullet X$ by $\Omega\mathbf{V}_\bullet L$ and R by $\mathbf{Z}$ and using 4.4(iv). In addition one has to verify that the realization of the simplicial space $\Omega\mathbf{V}_\bullet L$ is naturally weakly equivalent to ΩL. But this follows readily from an argument of Quillen [7] and the fact that the singular complex of the realization of $\Omega\mathbf{V}_\bullet L$ is naturally weakly equivalent to the diagonal of the bisimplicial set one obtains by applying the singular functor dimensionwise to the simplicial space $\Omega\mathbf{V}_\bullet L$.

This spectral sequence suggests the notion of

5.2 E-flat Π-algebras. A Π-algebra X is called *E-flat* if $(\mathbf{L}_p E)X = 0$ for $p > 0$.

An immediate consequence of 5.1 then is

5.3 A Milnor-Moore theorem. *If $L \in \mathbf{CW}_*$ is such that $\pi_* L$ is E-flat, then the canonical map $\mu' \colon E\pi_* L \to H_*\Omega L \in \mathbf{R}$ of 4.3 is an isomorphism.*

5.4 Examples. We now give examples of spaces $L \in \mathbf{CW}_*$ for which $\pi_* L$ is E-flat, so that the canonical map $\mu' \colon E\pi_* L \to H_*\Omega L$ of 4.3 is an isomorphism.

(i) *L is aspherical* (3.4(i)). This follows readily from the fact that dimensionwise application of the integral group ring functor to a homotopically discrete (i.e. $\pi_i = 0$ for $i > 0$) simplicial group, yields a homotopically discrete simplicial abelian group.

(ii) *L is a wedge of spheres.* Because then $\pi_* L$ is a free Π-algebra.

(iii) *L is simply connected and rational* (3.4(ii)). To prove this construct, by a slight modification of the procedure of [11, §2], a cellular simplicial resolution $M_\bullet$ of L in which each $M_i \in \mathbf{CW}_*$ has the homotopy type of a wedge of rational spheres of dimensions ≥ 2. Then (4.4(iv)) $E\pi_* M_\bullet$ is the dimensionwise enveloping algebra of the underlying simplicial Lie algebra of $\pi_* M_\bullet$. Moreover [7, §5] $E\pi_* M_\bullet$ admits a filtration for which the successive quotients are dimensionwise functors on $\pi_* M_\bullet$, considered as a simplicial rational vector space. As, by construction, $\pi_* M_\bullet$ is homotopically discrete, so is therefore [2] $E\pi_* M_\bullet$. The desired result now follows from the fact that [3] the spectral sequence of the bisimplicial ring $E\mathbf{V}_\bullet M_\bullet$ (where $\mathbf{V}_\bullet$ is as in [10, §2]) together with the E-flatness (in view of (ii) and 4.5) of the M_i ($i \geq 0$), implies that $(\mathbf{L}_p E)\pi_* L = \pi_p(E\pi_* M_\bullet)$ for all $p \geq 0$.

(iv) *Many other such spaces* $L \in \mathbf{CW}_*$ can be constructed from the above ones, using

5.5 Lemma. *Let $* \to U \to W \to V \to *$ be a short exact sequence of Π-algebras such that*

(i) U and V are E-flat, and

(ii) $\mathrm{Tor}_1^{\mathbf{Z}}(EV, EU) = 0$.

Then W is E-flat.

In view of the following proposition, which readily follows from 4.6, it suffices to prove this lemma for simply connected U, V and W. This will be done in 6.3.

5.6 Proposition. *A Π-algebra X is E-flat if and only if its universal covering $\tilde{X}$ (2.3) is E-flat.*

6 An Eilenberg-Moore spectral sequence

In this final section we construct a first quadrant Eilenberg-Moore spectral sequence for enveloping rings, which we then use to prove Lemma 5.5.

First we recall from [9, II, §6] and [10] a few facts concerning

6.1 Derived functors of the enveloping ring functor E. (i) Let X be a Π-algebra and let $X_\bullet \to X$ be a *free simplicial resolution* of X, i.e. $X_\bullet$ is a free simplicial Π-algebra and the map $X_\bullet \to X$ induces isomorphisms $\pi_i X_\bullet \approx \pi_i X$ for all $i \geq 0$ (where X is considered as a "discrete" simplicial Π-algebra, so that $\pi_0 X = X$ and $\pi_i X = 0$ for $i > 0$). Then the abelian group $\pi_p E X_\bullet$ $(p \geq 0)$, obtained by dimensionwise application of the functor E to the simplicial Π-algebra $X_\bullet$ and then taking the p-th homotopy group, does *not* depend on the choice of the free simplicial resolution of X and is denoted by $(\mathbf{L}_p E)X$. Similarly, given a map $f\colon X \to Y \in \mathbf{\Pi}$**-al** and a *free simplicial resolution* of f, i.e. a commutative diagram of simplicial Π-algebras

$$\begin{array}{ccc} X_\bullet & \xrightarrow{f_\bullet} & Y_\bullet \\ \downarrow & & \downarrow \\ X & \xrightarrow{f} & Y \end{array}$$

in which the vertical maps are free simplicial resolutions of X and Y respectively, the map

$$\pi_p E f_\bullet\colon \pi_p E X_\bullet = (\mathbf{L}_p E)X \to \pi_p E Y_\bullet = (\mathbf{L}_p E)Y$$

does *not* depend on the choice of this free simplicial resolution and is denoted by $(\mathbf{L}E)f$. The resulting functor $\mathbf{L}_p E\colon \mathbf{\Pi}$**-al** $\to$ **abelian groups** is called the p-th *left derived functor of* E. One readily verifies that $\mathbf{L}_0 E = E$.

(ii) Given a Π-algebra X and a free simplicial resolution $X_\bullet \to X$ of X, we have that $EX_\bullet$ is a simplicial ring and its multiplication turns the graded abelian group $(\mathbf{L}_* E)X = \{(\mathbf{L}_p E)X\}_{p=0}^{\infty}$ into a *ring*, which again does *not* depend on the choice of the free simplicial resolution. Similarly, a map $f\colon X \to Y \in \mathbf{\Pi}$**-al** gives rise to a *ring homomorphism* $(\mathbf{L}_* E)f\colon (\mathbf{L}_* E)X \to (\mathbf{L}_* E)Y$.

(iii) Free simplicial resolutions of Π-algebras and maps between them always exist. For example, if $X \in \mathbf{\Pi}$**-al**, one has the *standard free simplicial resolution* $F_\bullet X \to X$, in which each $(F_\bullet X)_n$ $(n \geq 0)$ consists of the **Π**-algebra $F^{n+1}X$, obtained by $(n+1)$-fold application of the free **Π**-algebra functor F (i.e. the forgetful functor **Π-al** $\to$ ((≥ 1)-**graded pointed sets**) followed by its left adjoint) and in which the face and degeneracy maps and the map $F_\bullet X \to X$ are the obvious ones.

All this enables us to obtain

6.2 A first quadrant Eilenberg-Moore spectral sequence for enveloping rings. *Let* $* \to U \to W \to V \to *$ *be a short exact sequence of* **Π**-*algebras. Then the induced isomorphism* (4.7) $EW \otimes_{EU} \mathbf{Z} \approx EV$ *extends to a first quadrant spectral sequence* $\{E^r_{p,q}\}$ *which converges to* $(\mathbf{L}_* E)V$ *and for which*

$$E^2_{p,q} = \operatorname{Tor}_p^{(\mathbf{L}_\bullet E)U}((\mathbf{L}_* E)W, \mathbf{Z})_q.$$

Proof. The map $W \to V$ induces (see 6.1(iii)) a map $F_\bullet W \to F_\bullet V$ of simplicial **Π**-algebras. Let $F'_\bullet U$ be its kernel. Then it is not difficult to show (using 2.5(v)) that

(i) the induced map $F'_\bullet U \to U$ is a free simplicial resolution of U, and

(ii) $EF_\bullet W$ is a free right simplicial module over $EF'_\bullet U$. As (4.7) the map $F_\bullet W \to F_\bullet V$ induces an isomorphism $EF_\bullet W \otimes_{EF'_\bullet U} \mathbf{Z} \approx EF_\bullet V$, application of Quillen's second Kunneth spectral sequence [9, II, §6] now yields the desired result. ∎

We now use 6.2 to give a

6.3 Proof of Lemma 5.5 (modulo Proposition 6.4). In view of 5.6 we may assume that U, V and W are simply connected. Moreover the E-flatness of U implies that $E^2_{p,q} = \operatorname{Tor}_p^{EU}((\mathbf{L}_q E)W, \mathbf{Z})$. As (4.7) $E^2_{0,0} = EW \otimes_{EU} Z \approx EV$, the convergence of the spectral sequence to EV (because V is E-flat), together with Proposition 6.4, successively implies that

$$E^2_{1,0} = \operatorname{Tor}_1^{EU}(EW, Z) = 0,$$
$$E^2_{p,0} = 0, \quad \text{for } p > 0,$$
$$E^2_{0,1} = \operatorname{Tor}_0^{EU}((\mathbf{L}_1 E)W, Z) = 0 \quad \text{and hence} \quad (\mathbf{L}_1 E)W = 0,$$
$$E^2_{p,1} = 0, \quad \text{for } p > 0,$$

and this argument can now be repeated inductively to show that $(\mathbf{L}_p E)W = 0$ for all $p > 0$, i.e. that W is E-flat. It thus remains to prove

6.4 Proposition. *Let* $R \in \mathbf{R}$ *be connected, i.e.* $R_0 = \mathbf{Z}$, *and let* M *be a non-negatively graded right* R-*module.*

(i) *If* $\operatorname{Tor}_0^R(M, \mathbf{Z}) = M \otimes_R \mathbf{Z} = 0$, *then* $M = 0$.

(ii) *If* $\operatorname{Tor}_1^R(M, \mathbf{Z}) = 0 = \operatorname{Tor}_1^{\mathbf{Z}}(M \otimes_R \mathbf{Z}, R)$, *then* $\operatorname{Tor}_p^R(M, \mathbf{Z}) = 0$ *for* $p > 1$.

Proof. (i) This follows immediately from the fact that the augmentation ideal $I \subset R$ contains only elements of positive degree.

(ii) Let $q \geq 1$ and assume that we already have shown that $\mathrm{Tor}_p^R(M, \mathbf{Z}) = 0$ for $q \geq p \geq 1$. Then it suffices to show that $\mathrm{Tor}_{q+1}^R(M, \mathbf{Z}) = 0$. To do this let $I_s \subset I$ consist of the elements of degree $\geq s$. A short exact sequence $0 \to F_1 \to F_0 \to I_s/I_{s+1} \to 0$ with F_0 and F_1 free abelian, then gives rise to a long exact ladder

$$\begin{array}{ccccccc} \cdots \to & \mathrm{Tor}_p^R(M, F_1) & \to & \mathrm{Tor}_p^R(M, F_0) & \to & \mathrm{Tor}_p^R(M, I_s/I_{s+1}) & \to \cdots \\ & \downarrow & & \downarrow & & \downarrow & \\ \cdots \to & \mathrm{Tor}_p^{\mathbf{Z}}(M \otimes_R \mathbf{Z}, F_1) & \to & \mathrm{Tor}_p^{\mathbf{Z}}(M \otimes_R \mathbf{Z}, F_0) & \to & \mathrm{Tor}_p^{\mathbf{Z}}(M \otimes_R \mathbf{Z}, I_s/I_{s+1}) & \to \cdots \end{array}$$

in which, for $q \geq p \geq 0$, the vertical maps involving F_0 and F_1 are readily verified to be isomorphisms. Consequently, by the five lemma, $\mathrm{Tor}_q^R(M, I_s/I_{s+1}) = 0$ for all $s > 0$ and hence $\mathrm{Tor}_q^R(M, I_s/I_{s+k}) = 0$ for all $s, k > 0$. Passing to the limit, one gets that $\mathrm{Tor}_q^R(M, I) = 0$ and hence, in view of the short exact sequence $0 \to I \to R \to \mathbf{Z} \to 0$, that $\mathrm{Tor}_{q+1}^R(M, \mathbf{Z}) = 0$.

References

[1] D.A. Blanc: A Hurewicz spectral sequence for homology. To appear.

[2] A. Dold: Homology of symmetric products and other functors of complexes. *Ann. of Math.* 68 (1958), 54–80.

[3] A. Dold and D. Puppe: Homologie nicht-additiver Functoren; Anwendungen. *Ann. Inst. Fourier* 11 (1961), 201–312.

[4] W.G. Dywer and D.M. Kan: Homology and cohomology of Π-algebras. To appear.

[5] T. Ganea: A generalization of the homology and the homotopy suspension. *Comment. Math. Helv.* 39 (1965), 295–322.

[6] J.W. Milnor: On the construction FK. *London Math. Soc. Lecture Note Ser. 4, pp. 119–136*, 1972.

[7] J.W. Milnor and J.C. Moore: On the structure of Hopf algebras. *Ann. of Math.* 81 (1965), 211–264.

[8] D.G. Quillen: Spectral sequences of a double semi-simplicial group. *Topology* 5 (1966), 155-156.

[9] D.G. Quillen: Homotopical algebra. Lecture Notes in Math. 43, Springer, 1967.

[10] D.G. Quillen: On the (co-)homology of commutative rings. *Proc. Sympos. Pure Math.* 17 (1970), 65–87.

[11] C.R. Stover: A van Kampen spectral sequence for higher homotopy groups. To appear in *Topology*.

[12] G.W. Whitehead: Elements of Homotopy Theory. Graduate Texts in Math. 61, Springer, 1978.

Cohomologie des groupes symétriques et application de Quillen

J. H. Gunawardena, J. Lannes, S. Zarati

1 Introduction

Soient G un groupe de Lie compact et BG son espace classifiant. Dans [Q1] Quillen considère l'application naturelle

$$q_G: H^*(BG, \mathbb{F}_2) \longrightarrow \varprojlim_{\mathcal{C}(G)} H^*(BV; \mathbb{F}_2),$$

$\mathcal{C}(G)$ désignant la catégorie dont les objets sont les 2-sous-groupes abéliens élémentaires V de G (rappelons qu'un 2-groupe abélien élémentaire est un groupe isomorphe à $(\mathbb{Z}/2)^d$ pour un certain entier $d \geq 0$) et dont les morphismes sont les applications linéaires $u: V \to V$ qui sont restrictions de conjugaisons dans G. Quillen montre que l'application q_G, que nous appellerons l'application de Quillen, vérifie:

1.1 Théorème. (a) *Tout élément du noyau de q_G est nilpotent.*

(b) *Pour tout élément y de $\varprojlim_{\mathcal{C}(G)}$ il existe un entier $r \geq 0$ tel que y^{2^r} appartient à l'image de q_G.*

Le but de cette note est de démontrer le Théorème suivant:

1.2 Théorème. *Soient G un groupe de Lie compact, S_n le groupe symétrique de degré n, $n \geq 1$, et $S_n \wr G$ le produit en couronne de S_n et de G. Si l'application de Quillen est un isomorphisme pour le groupe G, alors il en est de même pour le groupe $S_n \wr G$. En particulier l'application q_{S_n} est un isomorphisme.*

Il est à noter que l'énoncé ci-dessus est vérifié avec monomorphisme à la place d'isomorphisme; c'est alors un cas particulier de la Proposition 3.4 de [Q2].

Voici une description de notre méthode de démonstration du Théorème 1.2. Soit A l'algèbre de Steenrod modulo 2. L'application de Quillen est en fait un isomorphisme de A-algèbres instables (la cohomologie modulo 2 d'un espace est le type même d'une telle algèbre) et le Théorème 1.2 s'exprime à l'aide de l'homomorphisme de $\mathbb{F}_2$-algèbres commutatives sous-jacent. Notre idée est de linéariser le problème: le Théorème 1.2

s'exprime tout aussi bien à l'aide de l'homomorphisme de A-modules instables sous-jacent à q_G. En employant des techniques élémentaires d'algèbre homologique dans la catégorie abélienne des A-modules instables, nous dégageons une propriété du A-module instable $H^*(BG;\mathbb{F}_2)$, être "Nil-fermé", qui est nécessaire et suffisante pour que q_G soit un isomorphisme et nous montrons que si $H^*(BG;\mathbb{F}_2)$ est Nil-fermé, alors il en est de même pour $H^*(B(S_n \wr G);\mathbb{F}_2)$.

Le plan du papier est le suivant. Dans les paragraphes 2 à 4, nous introduisons la notion de A-module instable Nil-fermé et nous en donnons quelques propriétés. Le cinquième paragraphe est consacré à la démonstration du Théorème 1.2 à l'aide de cette notion. Dans le sixième paragraphe nous montrons que ce théorème n'est plus vrai si l'on remplace $H^*(\ ;\mathbb{F}_2)$ par $H^*(\ ;\mathbb{F}_p)$, p désignant un nombre premier impair.

J. Lannes et S. Zarati remercient L. Smith et tous les membres de la SFB à Götttingen pour leur hospitalité durant l'été 1986. Les trois auteurs remercient le Centre de Mathématiques de l'Ecole Polytechnique qui a facilité leur collaboration.

2 Définition et notations

Soit A l'algèbre de Steenrod modulo 2. Rappelons qu'un A-module M est dit instable si $Sq^i x = 0$ pour tout x de M et tout $i > |x|$, où $|x|$ désigne le degré de x. En particulier M est nul en degré strictement négatif. On note $\mathcal{U}$ la catégorie dont les objets sont les A-modules instables et dont les morphismes sont les applications A-linéaires de degré zéro.

Soit X un espace; la cohomologie modulo 2 de X, notée $H^*(X)$ dans les paragraphes 2 à 5, est non seulement un A-module instable, mais aussi une A-algèbre instable. Rappelons qu'une *A-algèbre instable* est un objet K qui possède les propriétés suivantes.

($\mathcal{K}1$) K est une algèbre graduée sur le corps $\mathbb{F}_2$ et, en plus, est associative commutative et unitaire;

($\mathcal{K}2$) K est un A-module instable et le produit: $K \otimes K \to K$ est A-linéaire;

($\mathcal{K}3$) Le "cup-carré" dans K coïncide avec l'application $x \mapsto Sq^{|x|}x$, $x \in K$.

Les A-algèbres instables sont les objets d'une catégorie, notée $\mathcal{K}$, dont les morphismes sont les applications de degré zéro, A-linéaires, compatibles avec produit et unité.

La propriété ($\mathcal{K}3$) d'une A-algèbre instable rappelée ci-dessus justifie les notations et la définition ci-dessous. Soit M un A-module et x un élément de M. On pose $Sq^{|x|}x = x^2$; on note $x \mapsto x^{2^r}$ la r-ième itérée de l'application $x \mapsto x^2$.

2.1 Définition. *On dit qu'un A-module instable N est nilpotent si pour tout élément x de N il existe un entier $r \geq 0$ tel que $x^{2^r} = 0$.*

On peut énoncer le Théorème 1.1 uniquement en termes de A-modules instables.

2.2 Théorème. *Soit G un groupe de Lie compact. Le noyau et le conoyau de l'homomorphisme de A-modules instables*

$$q_G: H^*(BG) \longrightarrow \lim_{\overleftarrow{\mathcal{C}(G)}} H^*(BV)$$

sont nilpotents.

2.3 Définition. *On dit qu'un A-module instable M est réduit si "l'élévation au carré": $M \to M$, $x \mapsto x^2$ est injective.*

Il est clair que M est réduit si et seulement si $\mathrm{Hom}_{\mathcal{U}}(N, M) = 0$ pour tout A-module nilpotent N.

Rappelons qu'on appelle $\mathcal{U}$-injectif un objet injectif de la catégorie $\mathcal{U}$. Dans [LZ1] on montre:

2.4 Proposition. *Tout A-module instable réduit se plonge dans un $\mathcal{U}$-injectif réduit.*

2.5 Définition. *On dit qu'un A-module instable M est Nil-fermé si* $\mathrm{Ext}^i_{\mathcal{U}}(N, M) = 0$ *pour $i = 0, 1$ et pour tout A-module instable nilpotent N.*

Il est clair que tout $\mathcal{U}$-injectif réduit est Nil-fermé (la réciproque est largement fausse).

2.6 Remarques. Si l'on note Nil la sous-catégorie pleine de $\mathcal{U}$ dont les objets sont les A-modules instables nilpotents, alors la terminologie Nil-fermé est un cas particulier de celle employée par Gabriel dans [G]. Dans [LZ1] les Nil-fermés sont appelés Δ-modules par référence à la "condition Δ" introduite par Adams-Wilkerson dans [AW] (voir aussi [SS]).

3 Quelques propriétés des Nil-fermés

Les propositions suivantes sont conséquence directe de 2.3 et 2.5.

3.1 Proposition. (a) *Soit $0 \to M \to M' \to M''$ une suite exacte dans $\mathcal{U}$. Si M' est Nil-fermé et M'' réduit, alors M est Nil-fermé.*

(b) *Soit $0 \to M' \to M \to M'' \to 0$ une suite exacte dans $\mathcal{U}$. Si M' et M'' sont Nil-fermés, alors il en est de même pour M.*

3.2 Proposition. *Soit $f: M_1 \to M_2$ un homomorphisme de A-modules instables. On suppose que $\ker f$ et $\operatorname{coker} f$ sont nilpotents et que M_2 est Nil-fermé. Alors, les deux conditions suivantes sont équivalentes:*

(i) *f est un isomorphisme;*

(ii) *M_1 est Nil-fermé.*

La condition ($\mathcal{K}3$) justifie à nouveau l'introduction de la terminologie suivante. Soit $M \subset M'$ une inclusion de A-modules instables réduits ; nous disons que M est *quadratiquement clos* dans M' si tout élément de M qui est un carré dans M', est aussi un carré dans M, c'est-à-dire si le quotient M'/M est réduit. Nous disons qu'un A-module instable réduit M est quadratiquement clos s'il est quadratiquement clos dans tout réduit qui le contient. La proposition suivante découle des Propositions 2.4 et 3.1.

3.3 Proposition. *Soit M un A-module instable réduit. Les propriétés suivantes sont équivalentes:*

(i) *M est Nil-fermé;*

(ii) *M est quadratiquement clos;*

(iii) *M est quadratiquement clos dans un $\mathcal{U}$-injectif réduit;*

(iv) *M est quadratiquement clos dans un Nil-fermé.*

Soient M_1 et M_2 deux A-modules instables ; on note $M_1 \otimes M_2$ le produit tensoriel de M_1 et de M_2 sur $\mathbf{F}_2$ muni de l'action "diagonale" de A définie à l'aide de la formule de Cartan.

3.4 Proposition. *Soient M_1 et M_2 deux Nil-fermés, alors le produit tensoriel $M_1 \otimes M_2$ est Nil-fermé.*

Démonstration. Il résulte des Propositions 2.4 et 3.1(a) qu'un module instable M est Nil-fermé si et seulement s'il admet un début de résolution injective $0 \to M \to K_0 \to K_1$ avec K_i un $\mathcal{U}$-injectif réduit, $i = 0, 1$. On montre dans [LZ1] que le produit tensoriel de deux $\mathcal{U}$-injectifs réduits est encore $\mathcal{U}$-injectif (réduit!): d'où la proposition. ∎

4 Exemples de A-modules instables Nil-fermés

4.1 Lemme. *Le A-module instable $H^*(B\mathbf{Z}/2)$ est Nil-fermé.*

Démonstration. En tant que $\mathbf{F}_2$-algèbre graduée, $H^*(B\mathbf{Z}/2)$ est librement engendrée par un générateur de degré un. Le A-module instable $H^*(B\mathbf{Z}/2)$ est donc réduit et quadratiquement clos (voir ci-dessus le paragraphe qui précède la Proposition 3.3). ∎

4.2 Soit V un 2-groupe abélien élémentaire ($V \approx (\mathbf{Z}/2)^d$, $d \geq 0$). Posons $P = H^*(B\mathbf{Z}/2)$. D'après la Proposition 3.4 $H^*(BV)$, qui est isomorphe au produit tensoriel $P \otimes \cdots \otimes P$, d fois, est Nil-fermé. On a en fait un résultat beaucoup plus fort: $H^*(BV)$ est un objet injectif dans la catégorie $\mathcal{U}$ [C], [M] (voir aussi [LZ1]).

4.3 Tout produit de Nil-fermés est Nil-fermé. Plus généralement, d'après 3.1(a), toute limite inverse de Nil-fermés est encore Nil-fermée. En particulier la limite inverse $\varprojlim_{\mathcal{C}(G)} H^*(BV)$ qui apparaît dans le Théorème 1.1 est un A-module instable Nil-fermé.

4.4 Tout $\mathcal{U}$-injectif réduit se plonge dans un produit $\prod_\alpha H^*(BV_\alpha)$, où V_α désigne une famille de 2-groupes abéliens élémentaires (voir [LS]). Un A-module instable M est donc Nil-fermé si et seulement s'il admet un début de résolution injective de la forme:

$$0 \longrightarrow M \longrightarrow \prod_\alpha H^*(BV_\alpha) \longrightarrow \prod_\beta H^*(BV_\beta).$$

Nous avons tout fait (voir 3.2 et 4.3) pour pouvoir énoncer:

4.5 **Proposition.** *Soit G un groupe de Lie compact. L'application de Quillen*

$$q_G : H^*(BG) \longrightarrow \varprojlim_{\mathcal{C}(G)} H^*(BV)$$

est un isomorphisme si et seulement si le A-module instable $H^*(BG)$ est Nil-fermé.

5 Démonstration du Théorème 1.2

D'après la Proposition 4.5 il faut montrer que si $H^*(BG)$ est Nil-fermé, alors il en est de même pour $H^*(B(S_n \wr G))$. Rappelons que l'on peut obtenir un espace classifiant du groupe $S_n \wr G$ à partir d'un espace classifiant du groupe G de la manière suivante. Soient n un entier ≥ 1 et X un espace. On désigne par ES_n un espace contractile sur lequel S_n opère librement, et on note $S_n X$ le quotient de $ES_n \times X^n$ par l'action diagonale de S_n, le groupe S_n opérant sur X^n par permutation des facteurs. L'espace $S_n(BG)$ est un espace classifiant pour le groupe $S_n \wr G$. Il nous suffit donc de montrer:

5.1 **Proposition.** *Soit X un espace. Si le A-module instable H^*X est Nil-fermé, alors il en est de même pour le A-module instable $H^*(S_n X)$.*

La démonstration de la Proposition 5.1 se fait en deux étapes. Dans la première étape, nous traitons le cas $n = 2$ (le cas $n = 1$ est trivial!) et dans la seconde, le cas général.

5.2 **Démonstration de la Proposition 5.1, pour n=2.** On a la suite exacte fonctorielle de A-modules instables [Mi],[Z]:

$$0 \longrightarrow \Lambda^2 H^*X \longrightarrow H^*(S_2 X) \longrightarrow R_1 H^*X \longrightarrow 0,$$

où $\bigwedge^2 H^*X$ est la deuxième puissance extérieure de H^*X et R_1H^*X l'image de $\Delta^*: H^*S_2X \to H^*BS_2 \otimes H^*X$, où $\Delta: BS_2 \times X \to S_2X$ désigne une "diagonale de Steenrod". D'après la Proposition 3.1(b), il nous faut montrer que $\bigwedge^2 H^*X$ et R_1H^*X sont Nil-fermés dès que H^*X l'est.

5.3 Lemme. *Soit M un A-module instable. Si M est Nil-fermé, alors il en est de même pour $\bigwedge^2 M$.*

Démonstration. On a la suite exacte des A-modules instables:

$$0 \longrightarrow \bigwedge^2 M \longrightarrow (M \otimes M)^{S_2} \longrightarrow \Phi M \longrightarrow 0 .$$

Ci-dessus ΦM est le A-module instable défini par [BK]:

$$(\Phi M)^n = \begin{cases} M^{n/2} & \text{si } n \equiv 0\,(2) \\ 0 & \text{si } n \equiv 1\,(2) \end{cases} \quad \text{et} \quad Sq^i \Phi x = \begin{cases} \Phi(Sq^{i/2}x) & \text{si } i \equiv 0\,(2) \\ 0 & \text{si } i \equiv 1\,(2), \end{cases}$$

Φx désignant l'élément de ΦM représenté par l'élément x de M ; $(M \otimes M)^{S_2}$ est le sous-A-module de $M \otimes M$ formé des éléments invariants sous l'action du groupe S_2 (S_2 opérant sur $M \otimes M$ par permutation des facteurs). Le Lemme 5.3 découle de la Proposition 3.1(a) puisque les modules $(M \otimes M)^{S_2}$ et ΦM sont respectivement Nil-fermé et réduit dès que M est Nil-fermé. ∎

Soit M un A-module instable. On définit fonctoriellement un sous-A-module R_1M de $H^*BS_2 \otimes M$ qui coïncide, quand $M = H^*X$, avec le A-module instable R_1H^*X considéré précédemment [S] (voir aussi [LZ2]).

5.4 Lemme. *Soit M un A-module instable. Si M est Nil-fermé, alors il en est de même pour R_1M.*

Démonstration. Posons encore $P = H^*BS_2$. D'après 3.4 et 4.1 le A-module instable $P \otimes M$ est Nil-fermé. D'après 3.3, il nous faut montrer que R_1M est quadratiquement clos dans $P \otimes M$.

Avant cela introduisons la convention suivante. Rappelons que l'application $\Phi M \to M$, $\Phi x \mapsto x^2$ est A-linéaire pour tout A-module instable M. Lorsque M est réduit, nous identifions, via cette application, ΦM au sous-A-module de M formé des carrés.

Avec cette convention, ce qu'il nous faut montrer est l'égalité suivante:

$$(R_1M) \cap \Phi(P \otimes M) = \Phi(R_1M) .$$

Celle-ci découle des formules suivantes:

$$(R_1M) \cap (P \otimes \Phi M) = R_1(\Phi M),$$

$$R_1(\Phi M) \cap ((\Phi P) \otimes M) = \Phi(R_1M) .$$

La première résulte du Corollaire 4.2.5 de [LZ2] et la deuxième ne présente pas de difficulté. ∎

5.5 Démonstration de la Proposition 5.1 dans le cas général. Soit D_n un 2-sous-groupe de Sylow de S_n. On pose, comme au début du paragraphe 5, $D_nX = ED_n \times_{D_n} X^n$. En tant que A-module instable $H^*(S_nX)$ est facteur direct dans $H^*(D_nX)$. Il suffit donc de prouver que $H^*(D_nX)$ est Nil-fermé dès que H^*X l'est. Puisque l'application naturelle $H^*(S_nX) \longrightarrow H^*(D_nX)$ donnée par le transfert est un morphisme dans $\mathcal{U}$ et non pas dans $\mathcal{K}$, la méthode de "linéarisation" montre ici toute son utilité.

Quand $n = 2^q$, D_{2^q} est un produit en couronne itéré $D_{2^q} = S_2 \wr \ldots \wr S_2$ et on a, à homotopie près, $D_{2^q}X = S_2 \ldots S_2 X$. Dans ce cas on "itère" le cas $n = 2$ précédemment établi.

Quand $n = 2^{q_1} + \cdots + 2^{q_2}$, $q_1 < \ldots < q_r$, $D_nX = D_{2^{q_1}}X \times \ldots \times D_{2^{q_r}}X$ à homotopie près. Dans ce cas on utilise la Proposition 3.4 et le cas $n = 2^q$. ∎

6 Le cas p > 2

On ne peut dans le Théorème 1.2 remplacer les 2-sous-groupes abéliens élémentaires et la cohomologie mod 2 par les p-sous-groupes abéliens élémentaires et la cohomologie mod p, p désignant un nombre premier impair. Par exemple l'application de Quillen

$$q_{S_{p^s}} : H^*(BS_{p^s}; \mathbb{F}_p) \longrightarrow \lim_{\overleftarrow{\mathcal{C}(S_{p^s})}} H^*(BV; \mathbb{F}_p)$$

n'est pas en général un isomorphisme. Pour voir ceci, on pose $V_s = (\mathbb{Z}/p)^s$ et on note H^* la cohomologie mod p d'un groupe discret (en l'occurence fini).

On identifie V_s à un sous-groupe de S_{p^s} en le faisant opérer sur lui-même par translation et en numérotant ses points.

On note Q_s le produit

$$\prod_{u \in H^1(V_s) - \{0\}} \beta u,$$

(β désignant le Bockstein $H^1V_s \to H^2V_s$) et $(H^*V_s)^{\mathrm{Aut}\,V_s}$ le sous-espace de H^*V_s invariant par les automorphismes de V_s. Comme V_s est maximal parmi les p-sous-groupes abéliens élémentaires de S_{p^s}, l'application naturelle

$$\lim_{\overleftarrow{\mathcal{C}(S_{p^s})}} H^*V \to H^*V_s$$

possède une section sur le sous-espace $Q_s(H^*V_s)^{\mathrm{Aut}\,V_s}$. On voit donc que si $q_{S_{p^s}}$ est un isomorphisme, alors $Q_s(H^*V_s)^{\mathrm{Aut}\,V_s}$ est contenu dans l'image de la restriction: $H^*S_{p^s} \to H^*V_s$. Ceci n'est pas vrai en général pour $p > 2$ (voir [Mu]). Il en résulte que $q_{S_{p^s}}$ n'est pas en général un isomorphisme pour $p > 2$.

Références

[AW] Adams J.F and Wilkerson C.: Finite H-spaces and algebras over the Steenrod algebra. *Ann. of Math.* 111 (1980), 95–143.

[BK] Bousfield A.K and Kan D.M.: The homotopy spectral sequence of a space with coefficients in a ring. *Topology* 11 (1972), 79–106.

[C] Carlsson G.: G.B. Segal's Burnside ring conjecture for $(\mathbb{Z}/2)^k$. *Topology* 22 (1983), 83–103.

[G] Gabriel P.: Des catégories abéliennes. *Bull. Soc. Math. France* 90 (1962) 323–448.

[LZ1] Lannes J. et Zarati S.: Sur les $\mathcal{U}$-injectifs. *Ann. Sci. École Norm. Sup.* 19 (1986), 303–333.

[LZ2] Lannes J. et Zarati S.: Sur les foncteurs dérivés de la déstabilisation. *Math. Z.* 194 (1987), 25–59.

[LS] Lannes J. et Schwartz L.: Sur la structure des A-modules instables. *Topology.* A paraître.

[M] Miller H.: The Sullivan conjecture on maps from classifying spaces. *Ann. of Math.* 120 (1984), 39–87.

[Mi] Milgram R.J.: Unstable homotopy from the stable point of view. Lecture Notes in Math. 368, Springer, 1974.

[Mu] Mui H.: Modular invariant theory and cohomology algebras of symmetric groups. *J. Fac. Sci. Univ. Tokyo* 22 (1975), 319–371.

[Q1] Quillen D.: The spectrum of an equivariant cohomology ring. I. II. Ann. of Math. 94 (1971), 549–572, 573–602.

[Q2] Quillen D.: The Adams conjecture. *Topology* 10 (1971), 67–80.

[S] Singer W.M.: The construction of certain algebras over the Steenrod algebra. *J. Pure Appl. Algebra* 11 (1977), 53–59.

[SS] Smith L. and Switzer R.M.: Polynomial algebras over the Steenrod algebra, variation on a theorem of Adams and Wilkerson. *Proc. Edinburgh Math. Soc.* 27, Part 1 (1984), 11–19.

[Z] Zarati S.: Défaut de stabilité d'opérations cohomologiques. *Publications Mathématiques d'Orsay* 78-07 (1978).

Looping Massey-Peterson towers

J. R. Harper and H. R. Miller

Introduction

In this paper we begin a study of the mod p cohomology of function spaces by means of an Adams resolution of the target space. For simplicity and concreteness we will require the source space to be a sphere, so we are dealing with iterated loop spaces. We will use the "classical" unstable Adams towers constructed by Massey and Peterson [4] and Barcus [1]. This technique restricts us to consideration of target spaces X with "very nice" cohomology: $H^*(X) = U(M)$ for some unstable module M over the Steenrod algebra $\mathcal{A}$, where U is the Steenrod-Epstein enveloping unstable $\mathcal{A}$-algebra functor. The class of such spaces includes Stiefel manifolds over C and H (and over R too if $p = 2$), but excludes nontrivial wedges and suspensions.

Our method is simply to compute, to the extent possible, the cohomology of the spaces in the k-fold loop space of a Massey-Peterson tower. By construction, the cohomology of each space E_s in such a tower for X surjects to $H^*(X)$. This fails for the looped tower, and there results a filtration of $H^*(\Omega^k X)$ in the category of $\mathcal{A}$-Hopf algebras. Our main result determines the associated quotient of $\mathcal{A}$-Hopf algebras, under rather restrictive connectivity assumptions, in terms of certain homological functors on the category $\mathcal{U}$ of unstable $\mathcal{A}$-modules. To describe these, let $\Omega^k M$ denote the maximal unstable quotient of the $\mathcal{A}$-module $\Sigma^{-k} M$. This is a right-exact functor from $\mathcal{U}$ to $\mathcal{U}$, and has left-derived functors Ω^k_s.

Theorem. Let X be a simply connected finite complex such that $H^*(X) = U(M)$ for some $M \in \mathcal{U}$. Let b and t be integers such that $M^i = 0$ unless $b < i \leq t$, and assume that $k - 1 \leq b - p^{-2}t$. Then there is a natural filtration of $H^*(\Omega^k X)$ by $\mathcal{A}$-Hopf algebras A_s such that $A_{-1} = \mathsf{F}_p$, $\bigcup A_s = H^*(\Omega^k X)$, and

$$A_{s+1}//A_s \cong U(\Omega \Omega^{s+k}_{s+1} M).$$

A more precise version of this theorem is stated in Section 3 as Theorem 3.9.

We remark that, while examples (like 4.10 below) show that some restrictions on M are necessary, the filtration by images of $H^*(\Omega^k E_s)$ seems to satisfy 3.9 in much more general circumstances than we have been able to prove here. For example, computations of W.M. Singer [5] show that $H^*(\Omega^k S^{n+k})$ has this form for all $k \geq 0$ and $n \geq 1$, at least for $p = 2$. Indeed, the derived functors he computes are precisely the ones entering into 3.9 if M is a k-fold suspension in $\mathcal{U}$.

We begin by reviewing certain elementary facts about derived functors of Ω^k. In Sections 2 and 3 we give a self-contained account of the Massey-Peterson-Barcus theory, using the Eilenberg-Moore spectral sequence. Finally, in Section 4, we prove the theorem.

We adopt the following convention on primes. When the case $p = 2$ can be handled by the substitution $\beta = \mathrm{Sq}^1$, $P^i = \mathrm{Sq}^{2i}$, we make no further comment; only when something special happens for $p = 2$ do we take notice.

Most of the work on this paper was done before 1980. The intervening years have seen the exploitation by Jean Lannes and others of more sophisticated variants of this approach, but using source spaces with more convenient cohomology. We offer our apologies for the long delay in publication, and express our gratitude for the opportunity to submit it to these proceedings in honor of Ioan James.

1 Some algebra

We begin with the algebraic loop functor Ω and its derived functors [4,5]. For any unstable $\mathcal{A}$-module M, define a vector space ΦM by

$$(\Phi M)^{2pn} = M^{2n},$$
$$(\Phi M)^{2pn+2} = M^{2n+1},$$
$$(\Phi M)^i = 0 \quad \text{otherwise.}$$

Write $\bar{x}$ for the element of ΦM corresponding to $x \in M$. It is direct to check that

1.1 ΦM becomes an unstable $\mathcal{A}$-module if we declare

$$P^{pj}\bar{x} = \overline{P^j x},$$
$$P^{pj+1}\bar{x} = \overline{\beta P^j x} \quad \text{if } |x| \text{ is odd,}$$
$$\beta^\epsilon P^j \bar{x} = 0 \quad \text{otherwise.}$$

1.2 The map $\lambda: \Phi M \to M$ defined by

$$\lambda\bar{x} = \begin{cases} P^n x & \text{if } |x| = 2n, \\ \beta P^n x & \text{if } |x| = 2n+1, \end{cases}$$

is $\mathcal{A}$-linear, and its kernel and cokernel are suspensions in $\mathcal{U}$.

Define functors Ω and Ω_1 by means of the resulting natural exact sequence

$$0 \longrightarrow \Sigma\Omega_1 M \longrightarrow \Phi M \xrightarrow{\lambda} M \longrightarrow \Sigma\Omega M \longrightarrow 0. \tag{1.3}$$

Since Φ is exact and λ is monic on projectives, standard homological methods imply that Ω_1 is the first left derived functor of Ω and that Ω has no higher derived functors.

The functor $\Omega: \mathcal{U} \to \mathcal{U}$ is right adjoint to Σ. Since the Σ is exact, Ω carries projectives to projectives. Consequently there is a Grothendieck spectral sequence associated to the composite $\Omega^j \Omega^k = \Omega^{j+k}$:

$$\Omega^j_s \Omega^k_t M \overset{s}{\Longrightarrow} \Omega^{j+k}_{s+t} M. \tag{1.4}$$

In particular, with $j = 1$, we obtain the *Singer exact sequence* [5]

$$0 \longrightarrow \Omega\Omega^k_s M \longrightarrow \Omega^{k+1}_s M \longrightarrow \Omega_1 \Omega^k_{s-1} M \longrightarrow 0. \tag{1.5}$$

One sees also by induction that

$$\Omega^k_s M = 0 \quad \text{if } s > k, \quad \text{and} \tag{1.6}$$

$$\Omega^s_s M = (\Omega_1)^s M. \tag{1.7}$$

We will need estimates on the connectivity and coconnectivity of $\Omega^{s+k}_s M$. If N is a graded vector space let

$$\operatorname{conn} N = \min\{i: N_i \neq 0\} - 1$$
$$\operatorname{coconn} N = \max\{i: N_i \neq 0\}.$$

1.8 **Lemma.** *If $M \in \mathcal{U}$, then*

$$\operatorname{conn} \Omega^{s+k}_s M \geq p^s(\operatorname{conn} M - k),$$
$$\operatorname{coconn} \Omega^{s+k}_s M \leq p^s \operatorname{coconn} M - k.$$

Proof. From (1.3) we have

$$\operatorname{conn} \Omega_1 M \geq p \operatorname{conn} M,$$
$$\operatorname{coconn} \Omega_1 M \leq p \operatorname{coconn} M.$$

(If p is odd, both inequalities can be improved by dividing according to parity, and in any case the second can be improved by at least 1. But these suffice for our purposes.) (1.7) then gives the result if $k = 0$. If $s = 0$ the result is clear, and the rest follows by induction using the Singer sequence (1.5). Notice that the cokernel determines the connectivity and the kernel determines the coconnectivity. ∎

The above algebra is connected with the geometry by the contravariant functor $K(\,)$ which associates to any projective unstable $\mathcal{A}$-module F of finite type a generalized Eilenberg-MacLane space $K(F)$ with homotopy $\pi_t K(F) \cong \operatorname{Hom}_{\mathcal{A}}(F, \Sigma^t \mathbf{F}_p)$. Then $H^*(F) \cong U(F)$; and to a morphism $f: F' \to F$ we associate the unique H-map

$K(f): K(F) \to K(F')$ inducing $U(f)$ in cohomology. The functor K is in a certain sense adjoint to H^*: (spaces) $\to \mathcal{U}$: there is a natural bijection

$$[X, K(F)] \cong \mathrm{Hom}_{\mathcal{U}}(F, H^*(X)).$$

There is a natural homotopy equivalence $K(\Omega^k F) \simeq \Omega^k K(F)$, so $H^*(\Omega^k K(F)) \cong U(\Omega^k F)$. For more general spaces X with $H^*(X) = U(M)$, $H^*(\Omega^k X)$ is much larger than $U(\Omega^k M)$; the isomorphism in this case is, as we shall see, a reflection of the triviality of the higher derived functors of Ω^{k+s} on F.

2 The cohomology theory of induced fibrations

The basic homological tool in the study of pull backs of fibrations is the Eilenberg-Moore spectral sequence. Suppose we have a pull-back diagram

$$\begin{array}{ccc} E & \xrightarrow{h} & PB_0 \\ \downarrow p & & \downarrow p_0 \\ B & \xrightarrow{f} & B_0 \end{array} \tag{2.1}$$

with p_0 the path-loop fibration. Under certain restrictions, an analysis of (2.1) was carried out by Massey and Peterson [3] and by Barcus [1], using the Serre spectral sequence. In this section we offer a proof of a variant of their results, using the Eilenberg-Moore machinery. We state all our results and comment on them first, and then sketch proofs.

To express the result conveniently we need the following universal construction. Given an unstable $\mathcal{A}$-algebra R and a map $g: R \to G$ of unstable $\mathcal{A}$-modules, we seek the initial unstable $\mathcal{A}$-algebra B accepting an $\mathcal{A}$-linear map from G such that the composite $R \to B$ is an $\mathcal{A}$-algebra homomorphism. Write $\overline{U}_R(G)$ for this B; it is a reduced version of a construction denoted U_R by Massey and Peterson. It is easily seen that if $U(R) \to R$ is the unique $\mathcal{A}$-algebra map extending the identity map $R \to R$, then

$$R \otimes_{U(R)} U(G) \xrightarrow{\cong} \overline{U}_R(G).$$

From this we see:

2.2 Lemma. *If g is monic then so is $R \to \overline{U}_R(G)$; and the extension*

$$0 \longrightarrow R \longrightarrow G \longrightarrow Q \longrightarrow 0$$

in $\mathcal{U}$ induces an extension

$$\mathsf{F}_p \longrightarrow R \longrightarrow \overline{U}_R(G) \longrightarrow U(Q) \longrightarrow \mathsf{F}_p$$

of $\mathcal{A}$-algebras: $\overline{U}_R(G)$ is free over R, and $U(Q) = \overline{U}_R(G) \otimes_R \mathsf{F}_p = \overline{U}_R(G)//R$. ∎

2.3 Theorem. *Consider the fiber square*

$$\begin{array}{ccc} E & \longrightarrow & PB_0 \\ {\scriptstyle p}\downarrow & & \downarrow \\ B & \xrightarrow{f} & B_0. \end{array}$$

Assume that B and B_0 are simply connected, that $H^*(B)$ and $H^*(B_0)$ are of finite type, and:

(i) *$H^*(B)$ is a free module over* $\operatorname{Im} f^*$.

Assume further that there is given a sub $\mathcal{A}$-module Q of $H^*(B_0)$ such that

(ii) $\Omega_1 Q = 0$;

(iii) *$H^*(B_0)$ is a free module over $U(Q)$; and*

(iv) $\ker f^*$ *is the ideal generated by Q.*

Let $R = H^*(B) \otimes_{H^*(B_0)} \mathbf{F}_p$. There is then a short exact sequence in $\mathcal{U}$, the "fundamental sequence"

$$0 \longrightarrow R \longrightarrow G \longrightarrow \Omega Q \longrightarrow 0, \tag{2.4}$$

and an extension $i: G \to H^*(E)$ of the natural $\mathcal{A}$-algebra map $R \to H^*(E)$, such that the induced map

$$\overline{U}_R(G) \longrightarrow H^*(E)$$

is an isomorphism, and

$$\begin{array}{ccc} G & \longrightarrow & \Omega Q \\ \downarrow & & \downarrow{\scriptstyle \omega} \\ H^*(E) & \xrightarrow{\partial^*} & H^*(\Omega B_0) \end{array} \tag{2.5}$$

commutes. Here ∂ is the Barratt-Puppe boundary map and ω is the natural "suspension" map.

Thus R embeds in $H^*(E)$, so $R \cong \operatorname{Im} p^*$.

As a special case, take $B = *$, so $E = \Omega B_0$. The theorem yields the

2.6 Corollary. *Let X be a simply connected space such that $H^*(X)$ is of finite type and $H^*(X) = U(Q)$, $\Omega_1 Q = 0$. Then the suspension map induces an isomorphism*

$$U(\Omega Q) \xrightarrow{\cong} H^*(\Omega X).$$

■

If X is k-connected and $\Omega_1 \Omega^i Q = 0$ for all i with $0 \leq i < k$, then by induction we have

$$U(\Omega^k Q) \xrightarrow{\cong} H^*(\Omega^k X).$$

The Singer sequence (1.5) shows that this condition is in fact equivalent to requiring $\Omega_{s+1}^{s+k}Q = 0$ for all $s \geq 0$; so our main theorem holds in this case without dimension assumptions.

2.7 Example. Typically one obtains the conditions of Theorem 2.3 by starting with B_0 such that $H^*(B_0) = U(N)$ for some $N \in \mathcal{U}$ for which $\Omega_1 N = 0$; for instance, $B_0 = K(N)$ for N projective in $\mathcal{U}$. Any $Q \subseteq N$ then has $\Omega_1 Q = 0$ as well, since Ω_1 is left exact. Note that $U(N)$ is an $\mathcal{A}$-Hopf algebra, with $PU(N) = N$. Now suppose that we can give $H^*(B)$ a Hopf algebra structure such that f^* is a Hopf algebra map. Then condition (i) of Theorem 2.3 is automatic. Take

$$Q = N \cap \ker f^* = \ker Pf^*. \tag{2.8}$$

Then we obtain (ii) of Theorem 2.3 as noted; (iii) holds because $U(Q) \to H^*(B_0)$ is monic, being a Hopf algebra map which is monic on primitives; and (iv) holds since $U(Q)$ is the Hopf algebra kernel $\operatorname{Ker} f^* = H^*(B_0) \square_{H^*(B)} \mathbb{F}_p$.

The Hopf condition on f^* usually occurs in one of two ways:

(a) $H^*(B) = U(M)$ for some $M \in \mathcal{U}$, and the map f^* induced by $f: B \to B_0$ fits into a diagram

$$\begin{array}{ccc} U(M) & \xleftarrow{U(\phi)} & U(N) \\ \cong\downarrow & & \cong\downarrow \\ H^*(B) & \xleftarrow{f^*} & H^*(B_0) \end{array}$$

for an $\mathcal{A}$-module map ϕ.

(b) B and B_0 are homotopy-associative H-spaces such that the Hopf algebra structure on $H^*(B_0) = U(N)$ has N for primitives, and f is an H-map. If $B_0 = K(N)$, the standard H-structure on B_0 will do, and f corresponds to a primitive cohomology class in B.

2.9 Remark. In case (b) above,

$$\begin{array}{ccc} E & \longrightarrow & PB_0 \\ p\downarrow & & \downarrow \\ B & \xrightarrow{f} & B_0 \end{array}$$

is a Hopf fiber square. We have seen that $R \cong \operatorname{Im} p^*$. We can identify other elements of the situation as follows:

(i) The Hopf algebra kernel $\operatorname{Ker} f^* = H^*(B_0) \square_{H^*(B)} \mathbb{F}_p = U(Q)$.

(ii) $\operatorname{Im} f^* = \operatorname{Ker} p^*$.

(iii) $\operatorname{Ker} \partial^*$ has a description in terms of a fundamental sequence. The diagram (2.5) shows that ∂^* factors as

$$\begin{array}{ccc} H^*(E) & \xrightarrow{\quad\partial^*\quad} & H^*(\Omega B_0) \\ \cong\uparrow & & \uparrow\cong \\ \overline{U}_R(G) \xrightarrow{q} U(\Omega \ker Pf^*) & \xrightarrow{U(\Omega k)} & U(\Omega N) \end{array}$$

where $k\colon \ker Pf^* \to N$ is the inclusion. By Lemma 2.2 the map q is epic with Hopf algebra kernel R, but $U(\Omega k)$ is not generally monic; indeed, there is an exact sequence

$$0 \longrightarrow \Omega_1 \operatorname{Im} Pf^* \xrightarrow{\delta} \Omega \ker Pf^* \xrightarrow{\Omega k} \Omega N \longrightarrow \Omega \operatorname{Im} Pf^* \longrightarrow 0.$$

If we form the pull-back diagram

$$\begin{array}{ccccccccc} 0 & \longrightarrow & R & \longrightarrow & K & \longrightarrow & \Omega_1 \operatorname{Im} Pf^* & \longrightarrow & 0 \\ & & =\downarrow & & \downarrow & & \downarrow\delta & & \\ 0 & \longrightarrow & R & \longrightarrow & G & \longrightarrow & \Omega \ker Pf^* & \longrightarrow & 0 \end{array}$$

then we see that the top row is a fundamental sequence for

$$\operatorname{Ker} \partial^* \cong \overline{U}_R(K).$$

It will be important for us to compare the fundamental sequence for E against that for ΩE. For this purpose we will write $Q(f)$ for the choice of Q, and agree to take $Q(\Omega f) = \ker P(\Omega f)^*$ as in (2.8). Then the suspension $\omega\colon \Omega H^*(B_0) \to H^*(\Omega B_0)$ automatically maps ΩQf to $Q(\Omega f)$. We will also write $R(f) = H^*(B) \otimes_{H^*(B_0)} \mathbb{F}_p$, etc.

Theorem 2.3 has the

2.10 Addendum. *Suppose $f\colon B \to B_0$ and $Q(f)$ satisfy the conditions of Theorem 2.3, that B and B_0 are 2-connected, and that $\Omega_1 Q(\Omega f) = 0$. Then there is $\omega\colon \Omega G(f) \to G(\Omega f)$ such that*

$$\begin{array}{ccccccccc} & & \Omega R(f) & \longrightarrow & \Omega G(f) & \longrightarrow & \Omega^2 Q(f) & \longrightarrow & 0 \\ & & \omega\downarrow & & \omega\downarrow & & \downarrow & & \\ 0 & \longrightarrow & R(\Omega f) & \longrightarrow & G(\Omega f) & \longrightarrow & \Omega Q(\Omega f) & \longrightarrow & 0 \end{array}$$

and

$$\begin{array}{ccc} \Omega G(f) & \longrightarrow & \Omega H^*(E) \\ \omega\downarrow & & \omega\downarrow \\ G(\Omega f) & \longrightarrow & H^*(\Omega E) \end{array}$$

commute.

We turn to a sketch of proofs. Write E_0 for PB_0. Let E_{0T} be the mapping cylinder of the projection map $p_0 \colon E_0 \to B_0$. Then $E_0 \to B_0$ factors canonically as $E_0 \rightarrowtail E_{0T} \xrightarrow{p_0'} B_0$, and p_0' is a homotopy equivalence. Now $E_0 \to E_{0T}$ is a cofibration over B inducing a surjection in cohomology. (This is the general assumption made in [3] by Massey and Peterson; here it is trivial since $E_0 \simeq *$.) Moreover $H^*(E_{0T}) \cong H^*(B_0)$ is a projective $H^*(B_0)$-module. It follows that the following diagram is the beginning of an exact couple yielding the Eilenberg-Moore spectral sequence (cf. Smith [6]):

$$\begin{array}{ccccc} H^*(B \times_{B_0} E_0) & \xrightarrow{\delta} & H^*(B \times_{B_0} (E_{0T}, E_0)) & \xrightarrow{\delta} & \cdots \\ & \nwarrow \quad \swarrow & & \nwarrow \quad \swarrow & \\ & H^*(B \times_{B_0} E_{0T}) & & H^*(B \times_{B_0} B_0 \times (E_{0T}, E_0)). & \end{array}$$

Rewriting this, we have

$$\begin{array}{ccccc} H^*(E) & \xrightarrow{\delta} & H^*(E_T, E) & \xrightarrow{\delta} & \cdots \\ & {\scriptstyle p^*}\nwarrow \quad \swarrow & & {\scriptstyle p_1^*}\nwarrow \quad \swarrow & \\ & H^*(B) & & H^*(B) \otimes H^*(E_{0T}, E_0) & \\ & \| & & \| & \\ & E_1^0 & \xleftarrow{d_1} & E_1^{-1}. & \end{array} \tag{2.11}$$

Therefore, in the Eilenberg-Moore filtration of $H^*(E)$,

$$F^0 = \ker \delta = \operatorname{Im} p^*,$$
$$F^{-1} = \ker \delta^2 = \delta^{-1} \operatorname{Im} p_1^*.$$

If we identify $H^*(E_{0T}, E_0) \cong \bar{H}^*(B_0)$, then $d_1(b \otimes b_0) = b \cdot f^* b_0$. Since $Q \subset \ker f^*$, the map $\Sigma^{-1} Q \to E_1^{-1}$ sending b_0 to $1 \otimes b_0$ carries $\Sigma^{-1} Q$ to $\ker(d_1 | E_1^{-1})$. Since $E_\infty^{-1} \in \mathcal{U}$, there is a factorization

$$\begin{array}{ccc} \Sigma^{-1} Q & \twoheadrightarrow & \Omega Q \\ \downarrow & & \downarrow \\ \ker(d_1 | E_1^{-1}) & \twoheadrightarrow & E_\infty^{-1}. \end{array} \tag{2.12}$$

Thus we may form the pull-back

$$\begin{array}{ccccccccc} 0 & \longrightarrow & R & \longrightarrow & G & \longrightarrow & \Omega Q & \longrightarrow & 0 \\ & & \| & & \downarrow & & \downarrow & & \\ 0 & \longrightarrow & F^0 & \longrightarrow & F^{-1} & \longrightarrow & E_\infty^{-1} & \longrightarrow & 0. \end{array} \tag{2.13}$$

The top sequence here is the fundamental sequence. The composite $G \to F^{-1} \rightarrowtail H^*(E)$ is the map i in the statement of Theorem 2.3. The composite $R \to G \to H^*(E)$ is an $\mathcal{A}$-algebra homomorphism, induced by p^*: so we get a map $\bar{i}: \overline{U}_R(G) \to H^*(E)$. To show it is an isomorphism, we filter $\overline{U}_R(G)$. First filter $U(G)$ by setting $F^0 = \mathsf{F}_p$, $F^{-1} = \mathsf{F}_p \oplus Q$, and $F^{-n} = (F^{-1})^n$; thus $F^{-p} = U(Q)$. Then let $F^{-n}\overline{U}_R(G)$ be the inverse image of $F^{-n}U(Q)$ under the natural map. It is then clear that $\bar{i}$ is filtration preserving when $H^*(E)$ is given the Eilenberg-Moore filtration. We will show that $E_0^*\bar{i}$ is an isomorphism.

Before we carry out this computation, we prove the naturality statements (2.5) and Addendum 2.10. The boundary map $\partial: \Omega B_0 \to E$ extends to a map of squares

$$\begin{array}{ccccccc} \Omega B_0 & \longrightarrow & E_0 & & E & \longrightarrow & E_0 \\ \downarrow & & \downarrow & \longrightarrow & \downarrow & & \downarrow \\ * & \longrightarrow & B_0 & & B & \longrightarrow & B_0, \end{array}$$

so compatibility with ∂ follows by naturality of the Eilenberg-Moore spectral sequence.

As for 2.10, note that the cohomology suspension map ω is induced from a natural transformation $\Sigma\Omega \to \mathrm{id}$. The relevant portion of the exact couple for the square with classifying map f is induced in H^* by

$$\begin{array}{ccccc} E & & (E_T, E) & & \\ & \searrow \;\; \nearrow & & \searrow & \\ & E_T & & & B \times (E_{0T}, E_0). \end{array} \tag{2.14}$$

The corresponding diagram for Ωf is

$$\begin{array}{ccccc} \Omega E & & (\Omega E_T, \Omega E) & & \\ & \searrow \;\; \nearrow & & \searrow & \\ & \Omega E_T & & & \Omega B \times (\Omega E_{0T}, \Omega E_0). \end{array} \tag{2.15}$$

The suspension of (2.15) maps to (2.14), and this leads to the desired naturality.

We now compute

$$E_2^* = \mathrm{Tor}^*_{H^*(B_0)}(H^*(B), \mathsf{F}_p)$$

in the Eilenberg-Moore spectral sequence. Let $A = \mathrm{Im}\, f^*$. Since $H^*(B)$ is a free A-module, the extension spectral sequence

$$\mathrm{Tor}^*_A(H^*(B), \mathrm{Tor}^*_{H^*(B_0)}(A, \mathsf{F}_p)) \Rightarrow \mathrm{Tor}^*_{H^*(B_0)}(H^*(B), \mathsf{F}_p)$$

collapses. Since $H^*(B_0)$ is free over $U(Q)$ and $A \cong H^*(B_0) \otimes_{U(Q)} \mathsf{F}_p$,

$$\mathrm{Tor}^*_{H^*(B_0)}(A, \mathsf{F}_p) \cong \mathrm{Tor}^*_{U(Q)}(\mathsf{F}_p, \mathsf{F}_p).$$

There is thus an $\mathcal{A}$-algebra isomorphism

$$R \otimes \mathrm{Tor}^*_{U(Q)}(\mathsf{F}_p, \mathsf{F}_p) \xrightarrow{\cong} \mathrm{Tor}^*_{H^*(B_0)}(H^*(B), \mathsf{F}_p). \tag{2.16}$$

Suppose first $p = 2$. The assumption $\Omega_1 Q = 0$ implies that $U(Q)$ is polynomial with $\mathrm{Tor}^{-1}_{U(Q)}(\mathsf{F}_2, \mathsf{F}_2) \cong \Sigma\Omega Q$ as $\mathcal{A}$-modules, so $\mathrm{Tor}^*_{H^*(B_0)}(H^*(B), \mathsf{F}_p) \cong R \otimes E[\sigma^{-1}\Sigma\Omega Q]$. By multiplicativity, $E_2 = E_\infty$. The map $\bar{\imath}$ is thus an isomorphism in this case.

Now take $p > 2$. By 1.1, the sub vector space $\Phi^+ Q$ of ΦQ of elements of degree divisible by $2p$ forms a sub $\mathcal{A}$-module. Since $\Omega_1 Q = 0$ we have a commutative diagram with exact rows and columns

$$\begin{array}{ccccccccc}
 & & & & 0 & & 0 & & \\
 & & & & \downarrow & & \downarrow & & \\
 & & 0 & \longrightarrow & \Phi^+ Q & \longrightarrow & \Phi Q & \longrightarrow & \Phi^- Q \longrightarrow 0 \\
 & & & & \downarrow & & \downarrow \lambda & & \\
 & & & & Q & \xrightarrow{=} & Q & & \\
 & & & & \downarrow & & \downarrow & & \\
0 & \longrightarrow & \Phi^- Q & \xrightarrow{\lambda^-} & IQ & \longrightarrow & \Sigma\Omega Q & \longrightarrow & 0 \\
 & & & & \downarrow & & \downarrow & & \\
 & & & & 0 & & 0 & &
\end{array}$$

which defines IQ and $\Phi^- Q$ in $\mathcal{U}$. Then as an algebra $U(Q)$ is free-commutative and $\mathrm{Tor}^{-1}_{U(Q)}(\mathsf{F}_p, \mathsf{F}_p) = I(Q)$ as $\mathcal{A}$-modules. Therefore $\mathrm{Tor}^*_{U(Q)}(\mathsf{F}_p, \mathsf{F}_p)$ is the free algebra with divided powers generated by $I(Q)$. The pth divided power $\gamma_p \colon \mathrm{Tor}^{-1}_{U(Q)}(\mathsf{F}_p, \mathsf{F}_p) \to \mathrm{Tor}^{-p}_{U(Q)}(\mathsf{F}_p, \mathsf{F}_p)$ factors through as an embedding

$$\Sigma^{p-2}\Phi^- Q \longrightarrow \mathrm{Tor}^{-p}_{U(Q)}(\mathsf{F}_p, \mathsf{F}_p)$$

sending $\bar{x}$ to $\gamma_p[x]$. Under this correspondence, the differential d_{p-1} satisfies

$$d_{p-1}\gamma_n[x] = \gamma_{n-p}[x] \cdot [\lambda^- \bar{x}] \tag{2.17}$$

up to a unit, for $n \geq p$. This follows from the fact that the divided powers in Tor are natural, so we can use them to compute the effect on Tor of the map of squares

$$\begin{array}{ccc} E & \longrightarrow & E_0 \\ \downarrow & & \downarrow \\ B & \longrightarrow & B_0 \end{array} \quad \longrightarrow \quad \begin{array}{ccc} K_{2n} & \longrightarrow & PK_{2n+1} \\ \downarrow & & \downarrow \\ * & \longrightarrow & K_{2n+1} \end{array}$$

to the appropriate universal example; here $|x| = 2n+1$ and $K_m = K(\mathbf{Z}/p, m)$. There, (2.17) is known ([6], Prop. 4.4, page 85).

The effect of d_{p-1} is that E_p^* contains no generators in homological degree less than -1, so $E_p^* = E_\infty^*$; and $E_p^* = E^0\overline{U}_R(G)$, so as before we conclude that $\overline{U}_R(G) \cong H^*(E)$. ∎

3 Massey-Peterson towers

We review the construction of a tower of fibrations associated to a simply connected space X with $H^*(X) = U(M)$ and a projective resolution

$$M \xleftarrow{\epsilon} P_0 \xleftarrow{d_0} P_1 \xleftarrow{d_1} P_2 \longleftarrow \cdots \tag{3.1}$$

in $\mathcal{U}$. Both $H^*(X)$ and P_s will be assumed to be of finite type, and P_s should be zero in degree $d \leq s+1$.

The tower will have the form

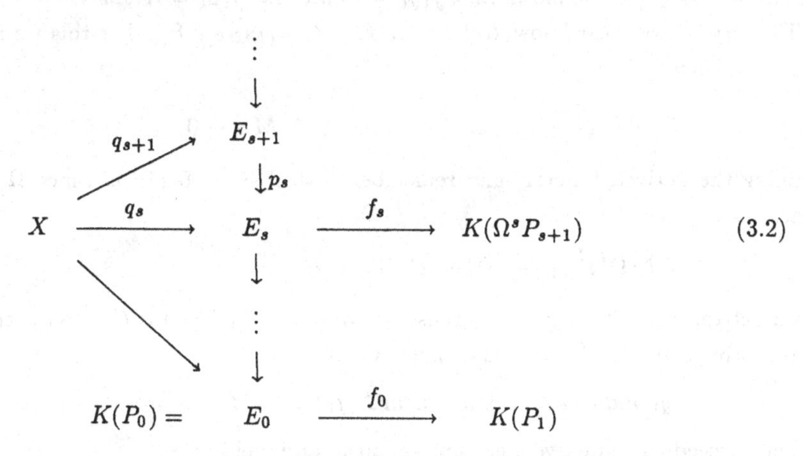

where p_s is the pull-back under the "k-invariant" f_s of the path-loop fibration; and if $\partial_s: K(\Omega^s P_s) \to E_s$ is the Barratt-Puppe boundary map, then

$$f_s \circ \partial_s \simeq K(\Omega^s d_s). \tag{3.3}$$

Begin by constructing $q_0: X \to K(P_0)$ such that $H^*(q_0) = U(\epsilon)$. Since $\epsilon d_0 = 0$, q_0 lifts to q_1. By Theorem 2.3 we have a diagram

$$\begin{array}{ccccccccc}
 & & & & & & \Omega P_2 & & \\
 & & & & & & \big\downarrow{\scriptstyle \Omega d_1'} & & \\
0 & \longrightarrow & R(f_0) & \longrightarrow & G(f_0) & \longrightarrow & \Omega \ker d_0 & \longrightarrow & 0 \\
 & & {\scriptstyle\cong}\big\downarrow & & \big\downarrow{\scriptstyle i} & & \big\downarrow & & \\
 & & H^*(X) & \xleftarrow{q_1^*} & H^*(E_1) & \xrightarrow{\partial_1^*} & H^*(K(\Omega P_1)) & &
\end{array} \tag{3.4}$$

in which $R(f_0) = H^*(K(P_0)) \otimes_{H^*(K(P_1))} \mathsf{F}_p \cong U(M) \cong H^*(X)$ and $\overline{U}_{R(f_0)}(G(f_0)) \xrightarrow{\cong} H^*(E_1)$. Thus $q_1^* i$ splits the fundamental sequence for E_1.

It follows that $H^*(E_1) \cong U(M \oplus \Omega \ker d_0)$ as $\mathcal{A}$-algebras, in such a way that the diagram

$$\begin{array}{ccccc} H^*(X) & \xleftarrow{q_1^*} & H^*(E_1) & \xrightarrow{\partial_1^*} & H^*(K(\Omega P_0)) \\ \uparrow & & \uparrow & & \uparrow \\ U(M) & \longleftarrow & U(M \oplus \Omega \ker d_0) & \longrightarrow & U(\Omega P_0) \end{array} \tag{3.5}$$

commutes. Here the bottom horizontal arrows are the obvious ones.

Let $\sigma: \Omega \ker d_0 \to M \oplus \Omega \ker d_0$ include the summand, and let $\Omega d_1': \Omega P_2 \to \Omega \ker d_0$ be induced from d_1. Then there exists a unique map $f_1: E_1 \to K(\Omega P_2)$ such that $U(\sigma \Omega d_1') = H^*(f_1)$. It follows that $f_1 q_1 = *$ and that $f_1 \partial_1 = K(\Omega d_1)$.

To carry on we must know $R(f_1) = H^*(E_1) \otimes_{H^*(K(\Omega P_2))} \mathsf{F}_p$. For this we note the exact sequence

$$\Omega^{s-1} P_{s+1} \xrightarrow{d_s'} \ker \Omega^{s-1} d_{s-1} \longrightarrow \Omega_s^{s-1} M \longrightarrow 0$$

defining the derived functor, and remember that $\Omega_s^{s-1} = 0$ (1.6). Since Ω is right exact,

$$\Omega d_s': \Omega^s P_{s+1} \longrightarrow \Omega \ker \Omega^{s-1} d_{s-1} \tag{3.6}$$

is surjective. Thus (taking $s = 1$) tensoring over $H^*(K(\Omega P_2)) = U(\Omega P_2)$ is the same as tensoring over $U(\Omega \ker d_0)$; and that gives us

3.7 *q_1 induces an isomorphism $R(f_1) \cong U(M)$.*

The succeeding inductive steps are identical, and yield

3.8 **Theorem** [4]. *There exists a tower (3.2) in which (3.3) holds and such that*

$$\begin{array}{ccccc} H^*(X) & \xleftarrow{q_s^*} & H^*(E_s) & \xrightarrow{\partial_s^*} & H^*(K(\Omega^s P_{s-1})) \\ \uparrow\cong & & \uparrow\cong & & \uparrow\cong \\ U(M) & \longleftarrow & U(M \oplus \Omega^s \ker d_{s-1}) & \longrightarrow & U(\Omega^s P_{s-1}) \end{array}$$

commutes.

Here we have slipped in the identification

$$\Omega^{s-1} \ker d_{s-1} \xrightarrow{\cong} \ker \Omega^{s-1} d_{s-1},$$

which follows from the exact sequence

$$0 \longrightarrow \Omega_s^{s-1} M \longrightarrow \Omega^{s-1} \ker d_{s-1} \longrightarrow \Omega^{s-1} P_s \xrightarrow{\Omega^{s-1} d_{s-1}} \Omega^{s-1} P_{s-1}$$

(see e.g. [2, page 93]) together with the fact that $\Omega_s^{s-1} M = 0$. A generalization of this identification, Lemma 4.5 below, is a key element to our approach and explains the form of the associated quotients in our main theorem, which we now restate in more precise form.

3.9 Theorem. *Let $M \in \mathcal{U}$ be of finite type, and let X be a simply connected space with $H^*(X) = U(M)$. Choose a projective resolution $M \longleftarrow P_*$ in $\mathcal{U}$ and construct the associated Massey-Peterson tower (3.2). For any*

$$k \leq \operatorname{conn} M - p^{-2} \operatorname{coconn} M + 1,$$

$$\operatorname{Im}(H^*(\Omega^k E_s) \longrightarrow H^*(\Omega^k X)) = \operatorname{Im}(H^*(\Omega^k E_s) \longrightarrow H^*(\Omega^k E_{s+1})),$$

and is independent of the resolution. If we call this Hopf algebra A_s, then there is a fundamental sequence giving rise to a Hopf algebra extension sequence over $\mathcal{A}$

$$\mathsf{F}_p \longrightarrow A_s \longrightarrow A_{s+1} \longrightarrow U(\Omega\Omega^{s+k}_{s+1} M) \longrightarrow \mathsf{F}_p.$$

4 Looping the tower: proof of the main theorem

Begin again as in Section 3, but assume X is $(k+1)$-connected and P_s is $(s+k)$-connected. We wish to compute $H^*(\Omega^k E_s)$ and the map $\Omega^k p_s^*: H^*(\Omega^k E_s) \rightarrow H^*(\Omega^k E_{s+1})$. Start with the fundamental sequence for $\Omega^k E_1$: by 2.10, we have a commutative diagram

$$\begin{array}{ccccccccc} 0 & \longrightarrow & \Omega^k U(M) & \longrightarrow & \Omega^k G(f_0) & \longrightarrow & \Omega^{k+1} \ker d_0 & \longrightarrow & 0 \\ & & \downarrow & & \downarrow \omega & & \downarrow \Omega\tau & & \\ 0 & \longrightarrow & U(\Omega^k M) & \longrightarrow & G(\Omega^k f_0) & \longrightarrow & \Omega \ker \Omega^k d_0 & \longrightarrow & 0. \end{array} \qquad (4.1)$$

The bottom left term, $R(\Omega^k f_0)$, is as displayed because Ω^k is right exact. Notice that $R(\Omega^k f_0) \rightarrow H^*(\Omega^k E_1)$ is a Hopf algebra map and that $PR(\Omega^k f_0) = \Omega^k M$ is zero in degrees greater than $\operatorname{coconn} M - k$.

We next state the inductive assumptions:

4.2 (i)$_s$ The fundamental sequence for $H^*(\Omega^k E_s)$ is given by the bottom row of the commutative diagram

$$\begin{array}{ccccccccc} 0 & \rightarrow & \Omega^k R(f_{s-1}) & \rightarrow & \Omega^k G(f_{s-1}) & \rightarrow & \Omega^{s+k} \ker d_{s-1} & \rightarrow & 0 \\ & & \downarrow & & \downarrow \omega & & \downarrow \Omega\tau & & \\ 0 & \rightarrow & R(\Omega^k f_{s-1}) & \rightarrow & G(\Omega^k f_{s-1}) & \rightarrow & \Omega \ker \Omega^{s+k-1} d_{s-1} & \rightarrow & 0 \end{array}$$

in which $\Omega\tau$ is the natural map.

(ii)$_s$ There is a Hopf algebra fundamental sequence for $R(\Omega^k f_{s-1})$ of the form

$$0 \longrightarrow R(\Omega^k f_{s-2}) \longrightarrow G_{s-1} \longrightarrow \Omega\Omega^{s+k-2}_{s-1} M \longrightarrow 0.$$

(iii)$_s$ $\operatorname{coconn} PR(\Omega^k f_{s-1}) \leq p^{s-1} \operatorname{coconn} M - k$.

Let $\sigma: \Omega^s \ker d_{s-1} \to G(f_{s-1})$ be the splitting (as in Section 3). The k-invariant $f_s: E_s \to K(\Omega^s P_{s+1})$ is defined as induced by the composite

$$\Omega^s P_{s+1} \xrightarrow{\Omega^s d_s'} \Omega^s \ker d_{s-1} \xrightarrow{\sigma} G(f_{s-1}) \longrightarrow H^*(E_s).$$

Since ω is compatible with the cohomology suspension by 2.10, it follows that the k-invariant $\Omega^k f_s$ is induced by the composite

$$\begin{array}{ccccc} \Omega^{s+k} P_{s+1} & \xrightarrow{\Omega^{s+k} d_s'} & \Omega^{s+k} \ker d_{s-1} & \xrightarrow{\Omega^k \sigma} & \Omega^k G(f_{s-1}) \\ & {\scriptstyle f} \searrow & & {\scriptstyle \hat{\sigma}} \searrow & \downarrow \omega \\ & & & & G(\Omega^k f_{s-1}) \rightarrowtail H^*(\Omega^k E_s). \end{array} \tag{4.3}$$

Write $\hat{\sigma} = \omega \circ \Omega^k \sigma$. We need to know $H^*(\Omega^k E_s)//\mathrm{Im}\, \Omega^k f_s^*$ and $\ker P(\Omega^k f_s)^*$, and for this it will suffice to find $\ker \hat{\sigma}$ and $\mathrm{coker}\, \hat{\sigma}$.

A diagram chase shows that $\hat{\sigma}$ is compatible with $\Omega\tau$, and there results a commutative diagram with exact rows which defines an important map "fat theta".

$$\begin{array}{ccccccccc} 0 & \longrightarrow & \ker \Omega\tau & \longrightarrow & \Omega^{s+k} \ker d_{s-1} & \longrightarrow & \mathrm{Im}\, \Omega\tau & \longrightarrow & 0 \\ & & \downarrow \Theta & & \downarrow \hat{\sigma} & & \Omega\tau \downarrow & & \\ 0 & \longrightarrow & R(\Omega^k f_{s-1}) & \longrightarrow & G(\Omega^k f_{s-1}) & \longrightarrow & \Omega \ker \Omega^{s+k-1} d_{s-1} & \longrightarrow & 0. \end{array} \tag{4.4}$$

We pause to record some information about τ:

4.5 Lemma. *There exists a natural commutative diagram in which each row and column is exact:*

$$\begin{array}{ccccccccc} & & 0 & & 0 & & 0 & & \\ & & \downarrow & & \downarrow & & \downarrow & & \\ 0 & \longrightarrow & \mathrm{Im}\, \Omega^{s+k} d_{s+1} & \longrightarrow & \ker \Omega^{s+k} d_s & \longrightarrow & \Omega^{s+k}_{s+1} M & \longrightarrow & 0 \\ & & = \downarrow & & \downarrow & & \downarrow & & \\ 0 & \longrightarrow & \mathrm{Im}\, \Omega^{s+k} d_{s+1} & \longrightarrow & \Omega^{s+k} P_{s+1} & \longrightarrow & \Omega^{s+k} \ker d_{s-1} & \longrightarrow & 0 \\ & & & & \downarrow \Omega\alpha & & \downarrow \Omega\tau & & \\ & & & & \Omega \ker \Omega^{s+k-1} d_{s-1} & \xrightarrow{=} & \Omega \ker \Omega^{s+k-1} d_{s-1} & & \\ & & & & \downarrow & & \downarrow & & \\ & & & & \Omega\Omega^{s+k-1}_s M & \xrightarrow{=} & \Omega\Omega^{s+k-1}_s M & & \\ & & & & \downarrow & & \downarrow & & \\ & & & & 0 & & 0\,. & & \end{array}$$

Proof. The top row defines $\Omega^{s+k}_{s+1} M$, and the second row follows from exactness of

$$P_{s+2} \xrightarrow{d_{s+1}} P_{s+1} \longrightarrow \ker d_{s-1} \longrightarrow 0$$

and right-exactness of Ω^{s+k}. The rows are thus exact. The map $\ker \Omega^{s+k} d_s \to \Omega^{s+k} P_{s+1}$ is the inclusion. The box to its left is clearly commutative, and this defines the vertical arrow to its right. We now construct the rest of the middle column and show it is exact; a diagram chase shows that this suffices.

The bottom three terms of the middle column are Ω applied to the exact sequence

$$\Omega^{s+k-1} P_{s+1} \xrightarrow{\alpha} \ker \Omega^{s+k-1} d_{s-1} \longrightarrow \Omega_s^{s+k-1} M \longrightarrow 0.$$

It remains to prove exactness at $\Omega^{s+k} P_{s+1}$. For this, consider the short exact sequence

$$\begin{array}{ccccccccc} & & \Omega^{s+k-1} P_{s+1} & & & & & & \\ & & \alpha \downarrow & \searrow^{\Omega^{s+k-1} d_s} & & & & & \\ 0 & \to & \ker \Omega^{s+k-1} d_{s-1} & \xrightarrow{i} & \Omega^{s+k-1} P_s & \longrightarrow & \operatorname{Im} \Omega^{s+k-1} d_{s-1} & \to & 0. \end{array}$$

Since $\operatorname{Im} \Omega^{s+k-1} d_{s-1}$ is a submodule of the projective $\Omega^{s+k-1} P_{s-1}$, and Ω_1 is left exact, Ωi is monic. Therefore $\ker \Omega\alpha$ and $\ker \Omega^{s+k} d_s$ coincide. ∎

Now by 2.10, $\omega: \Omega^k G(f_{s-1}) \to G(\Omega^k f_{s-1})$ is compatible with the cohomology suspension $\Omega^k H^*(E_s) \to H^*(\Omega^k E_s)$, which lands in the coalgebra primitives. Since $R(\Omega^k f_{s-1}) = H^*(\Omega^k E_{s-1})//\operatorname{Im}(\Omega^k f_{s-1}^*) \to H^*(\Omega^k E_s)$ is a Hopf algebra monomorphism, we conclude that Θ takes values in $PR(\Omega^k f_{s-1})$. By Lemmas 4.5 and 1.8, we know that

$$\operatorname{conn} \ker \Omega\tau \geq p^{s+1}(\operatorname{conn} M - k + 1).$$

By inductive assumption 4.2(iii)

$$\operatorname{coconn} PR(\Omega^k f_{s-1}) \leq p^{s-1} \operatorname{coconn} M - k.$$

The assumption

$$k - 1 \leq \operatorname{conn} M - p^{-2} \operatorname{coconn} M$$

of Theorem 3.9 lets us conclude that

$$\Theta = 0.$$

The serpent lemma applied to (4.4), together with Lemma 4.5, then gives an isomorphism

$$\Omega_{s+1}^{s+k} M \xrightarrow{\cong} \ker \hat{\sigma} \tag{4.6}$$

and a short exact sequence

$$0 \longrightarrow R(\Omega^k f_{s-1}) \longrightarrow \operatorname{coker} \hat{\sigma} \longrightarrow \Omega\Omega_s^{s+k-1} M \longrightarrow 0. \tag{4.7}$$

We can now verify the next stage of the inductive assumption 4.2. The third term in the fundamental sequence for $H^*(\Omega^k E_{s+1})$ is the algebraic loops of the kernel of (4.3). To find this kernel, consider the following diagram, whose top half coincides with the diagram in Lemma 4.5. The map f occurs in (4.3).

$$\begin{array}{ccccccccc}
 & & & & 0 & & 0 & & \\
 & & & & \downarrow & & \downarrow & & \\
0 & \longrightarrow & \operatorname{Im}\Omega^{s+k}d_{s+1} & \longrightarrow & \ker\Omega^{s+k}d_s & \longrightarrow & \Omega^{s+k}_{s+1}M & \longrightarrow & 0 \\
 & & \downarrow = & & \downarrow & & \downarrow & & \\
0 & \longrightarrow & \operatorname{Im}\Omega^{s+k}d_{s+1} & \longrightarrow & \Omega^{s+k}P_{s+1} & \longrightarrow & \Omega^{s+k}\ker d_{s-1} & \longrightarrow & 0 \\
 & & & & \downarrow f & & \hat{\sigma}\downarrow & & \\
 & & & & G(\Omega^k f_{s-1}) & \xrightarrow{=} & G(\Omega^k f_{s-1}) & & \\
 & & & & \downarrow & & \downarrow & & \\
 & & & & \operatorname{coker} f & \xrightarrow{=} & \operatorname{coker}\hat{\sigma} & & \\
 & & & & \downarrow & & \downarrow & & \\
 & & & & 0 & & 0. & &
\end{array} \tag{4.8}$$

Now the rows and the right column are known to be exact, so the middle column is too. We conclude that $4.2(\mathrm{i})_{s+1}$ holds.

To obtain $(\mathrm{ii})_{s+1}$ we establish a fundamental sequence for

$$R(\Omega^k f_s) = H^*(\Omega^k E_s) \otimes_{U(\Omega^{s+k}P_{s+1})} \mathsf{F}_p.$$

By (4.3), the k-invariant f_s^* factors through $U(\hat{\sigma})$ mapping to the second factor in

$$H^*(\Omega^k E_s) = R(\Omega^k f_{s-1}) \otimes_{UR(\Omega^k f_{s-1})} UG(\Omega^k f_{s-1}),$$

so by associativity of tensor product and right exactness of U,

$$R(\Omega^k f_s) \cong R(\Omega^k f_{s-1}) \otimes_{UR(\Omega^k f_{s-1})} U\operatorname{coker}\hat{\sigma}.$$

Thus (4.7) is a fundamental sequence for $R(\Omega^k f_s)$, establishing $(\mathrm{ii})_{s+1}$.

Assumption $4.2(\mathrm{iii})_{s+1}$ follows by applying P to the Hopf algebra short exact sequence

$$\mathsf{F}_p \longrightarrow R(\Omega^k f_{s-1}) \longrightarrow R(\Omega^k f_s) \longrightarrow U(\Omega\Omega_s^{s+k-1}M) \longrightarrow \mathsf{F}_p \tag{4.9}$$

associated to this fundamental sequence, and using Lemma 1.8.

This completes the proof of Theorem 3.9. The unfortunate restrictions on the coconnectivity of M are needed to kill fat theta in the above argument. A reduction in the estimate of the size of the module of primitives in $R(\Omega^k f_s)$ would lead to a better theorem; in homology one wants to guarantee the creation of squares. Examples of D. Kraines seem relevant here. However, some restrictions are necessary, as shown by the very simple

4.10 Example. Consider the two-stage Postnikov system

$$\begin{array}{ccc} E & \longrightarrow & * \\ \downarrow & & \downarrow \\ K(\mathbf{Z},2) & \xrightarrow{\mathrm{Sq}^2} & K(\mathbf{Z},4). \end{array}$$

Let $F'(n) = PH^*K(\mathbf{Z},n)$, so that $H^*K(\mathbf{Z},n) = UF'(n)$. Then the Massey-Peterson theory of Section 2 applies to compute $H^*(E)$, since $(\mathrm{Sq}^2)^* = U(\phi)$ where $\phi\colon F'(4) \to F'(2)$ sends ι_4 to $\mathrm{Sq}^2\iota_2$. The fundamental sequence

$$0 \longrightarrow E[\iota_2] \longrightarrow G \longrightarrow \Omega \ker\phi \longrightarrow 0$$

splits for degree reasons, so $H^*(E) = U(M)$ with

$$M = \langle \iota_2 \rangle \oplus \Omega \ker\phi.$$

Consider $H^*(\Omega E)$. The modules $\Omega\Omega_s^s(\Omega\ker\phi)$ are involved in Theorem 3.9. To compute them we have the short exact sequence

$$0 \longrightarrow \ker\phi \longrightarrow F'(4) \longrightarrow \langle \iota_4, \mathrm{Sq}^4\iota_4, \mathrm{Sq}^8\mathrm{Sq}^4\iota_4, \ldots \rangle \longrightarrow 0.$$

The right term has trivial Ω_1, so

$$0 \longrightarrow \Omega\ker\phi \longrightarrow F'(3) \longrightarrow \langle \iota_3 \rangle \longrightarrow 0$$

is still exact. Since Ω_1 is left exact and $\Omega_1 F'(n) = 0$, $\Omega_1\Omega\ker\phi = 0$. Thus

$$\Omega\Omega_s^s M = \langle \iota_{2^s} \rangle$$

for any $s \geq 1$, while

$$U(\Omega\ker\phi) = H^*(S^3\langle 3\rangle).$$

So our estimate of $H^*(\Omega E)$ resembles $H^*(\Omega S^2 \times S^3\langle 3\rangle)$.

But in fact $\Omega E = K(\mathbf{Z},1) \times K(\mathbf{Z},2)$, since $\Omega\mathrm{Sq}^2 \simeq *$. In understanding this example it may help to notice the square of fibrations

$$\begin{array}{ccccc} & & S^3\langle 3\rangle & = & S^3\langle 3\rangle \\ & & \downarrow & & \downarrow \\ S^1 & \longrightarrow & S^3 & \longrightarrow & S^2 \\ \| & & \downarrow & & \downarrow \\ S^1 & \longrightarrow & K(\mathbf{Z},3) & \longrightarrow & E \end{array}$$

which when looped back gives

$$\begin{array}{ccccc}
\Omega S^3 & \longrightarrow & \Omega S^2 & \longrightarrow & S^1 \\
\| & & \downarrow & & \| \\
K(\mathbb{Z},2) & \longrightarrow & \Omega E & \longrightarrow & S^1 \\
\downarrow & & \downarrow & & \\
S^3\langle 3\rangle & = & S^3\langle 3\rangle . & &
\end{array}$$

The horizontal sequence splits; we have picked up the vertical sequence, which does not.

References

[1] W.D. Barcus: On a theorem of Massey and Peterson. *Quart. J. Math. Oxford* 19 (1968), 33–41.

[2] H. Cartan and S. Eilenberg: Homological algebra. Princeton University Press, 1956.

[3] W.S. Massey and F.P. Peterson: The cohomology structure of certain fibre spaces. I. *Topology* 4 (1965), 47–65.

[4] W.S. Massey and F.P. Peterson: The mod 2 cohomology structure of certain fibre spaces. *Mem. Amer. Math. Soc.* 74 (1967).

[5] W.M. Singer: Iterated loop functors and the homology of the Steenrod algebra. *J. Pure Appl. Algebra* 11 (1977), 83–101.

[6] L. Smith: Lectures on the Eilenberg-Moore Spectral Sequence. Lecture Notes in Math. 134, Springer, 1970.

Characters and elliptic cohomology

Michael J. Hopkins

Introduction

The organization of stable homotopy theory in the large is detected by a family, $\{E_n\}_{n\geq 0}$, of complex oriented cohomology theories [R],[DHS],[Ho]. The first two are familiar: E_0 can be taken to be ordinary cohomology with coefficients in the real numbers, and E_1 is K-theory. These two are also associated with phenomena in geometry and analysis: ordinary cohomology with integration of forms via the de Rham theorem, and K-theory with the indices of elliptic operators via the index theorem. It is believed that there is a geometric interpretation of the rest of the $\{E_n\}$.

Under certain conditions [L2], the functor

$$E_*(\_) = R \underset{MU_*}{\otimes} MU_*(\_),$$

determined by a map

$$f : MU_* \longrightarrow R,$$

from the complex cobordism ring to a ring R, can be shown to be a cohomology theory. The $\{E_n\}$ are constructed in this way, from suitable maps

$$f_n : MU_* \longrightarrow R_n.$$

Each f_n can be thought of as a *genus* in the sense of Hirzebruch [H], and the search for a geometric interpretation of E_n breaks into two steps: first, to describe the new genus in terms of geometry and analysis, and second, to describe how the analytical genus changes when the manifold moves through a family. The elements of $E_n^*(X)$ can then be interpreted as families of analytical genera parameterized by X.

One difficulty with the first step is that the choice of f_n is not canonical. The ability of the cohomology theory E_n to detect the right phenomena in stable homotopy theory depends only on a weak invariant of the map f_n known as its *height*. For example, each of the cohomology theories

$$H^*(X; R) \qquad R \text{ a } \mathbb{Q}\text{-algebra},$$

detect stable homotopy theory modulo torsion. These cohomology theories can all be described by some sort of de Rham theorem [Ca], but it is cohomology with coefficients in the real numbers which is most naturally geometric. A prerequisite for a good choice of E_n, is therefore that the map f_n (and hence the ring R_n) have something to do with analysis.

On the other hand, something general can be said about the second step. To do this requires believing that the way a genus deforms in families is determined by an associated equivariant genus, which captures the effect of action by a group G, and that the equivariant genus amounts to describing the cohomology

$$E_n^*(BG).$$

The first of these beliefs is a fairly standard geometrical manipulation, whereas the second lies behind some interesting problems in algebraic topology [C]. The main result of [HKR] is a computation of $E_n^*(BG)$ (modulo conjecturally nonexistent torsion) for any choice of E_n, and for any finite group G. This description generalizes Atiyah's computation [A] of the K-theory of BG in terms of the representation ring of G, and will be described in the first part of this note.

So far the only values of n for which there is a natural 'analytical' choice of map

$$f_n : MU_* \longrightarrow R_n$$

are $n = 0, 1, 2$. The cases $n = 0$ (ordinary cohomology) and $n = 1$ (K-theory) have already been discussed. There is a natural height 2 genus with values in a certain ring of modular forms. It is known as the *elliptic genus*, and the associated cohomology theory, Ell^*, is called *elliptic cohomology* [LRS]. The elliptic genus was first described by Morava [M], and was later rediscovered, and made use of, by Ochanine [O].

Witten [W] has announced an analytical interpretation of the elliptic genus. This raises the possibility of finding a geometric interpretation of elliptic cohomology. Unfortunately, no one has, as of yet, succeeded in doing this.

The point of this note is to explain the main result of [HKR], and to describe how it signposts the possible geometric descriptions of the cohomology theories E_n in general, and elliptic cohomology in particular.

1 Formal groups

A (1-dimensional commutative) formal group law over a ring R is a formal power series

$$F(x, y) = x \underset{F}{+} y \in R[\![x, y]\!]$$

satisfying

$$(x \underset{F}{+} y) \underset{F}{+} z = x \underset{F}{+} (y \underset{F}{+} z),$$
$$x \underset{F}{+} y = y \underset{F}{+} x,$$
$$x \underset{F}{+} 0 = x.$$

Formal groups can also be described in terms of formal geometry. For a ring R let $\mathcal{C}_R$ be the category of *adic* R-algebras. Thus $\mathcal{C}_R$ consists of augmented R-algebras

$$A \xrightarrow{\varepsilon} R$$

with a nilpotent augmentation ideal $\mathfrak{m}$. The functor $D^n(A) = \mathfrak{m}^n$ is a model for affine n-space over R. Indeed, a map $D^n \to D^m$ associates to every n-tuple of nilpotent elements $\underline{a} = (a_1, \ldots, a_n)$ an m-tuple

$$\left(\sum b_{I_1} \underline{a}^{I_1}, \ldots, \sum b_{I_m} \underline{a}^{I_m}\right)$$
$$I_j = (i_{1,j}, \ldots, i_{n,j})$$
$$\underline{a}^{I_j} = a_1^{i_{1,j}} \cdots a_n^{i_{n,j}},$$

placing the set of natural transformations $D^n \to D^m$ in one to one correspondence with the set of m-tuples of formal power series

$$\left(\sum b_{I_1} \underline{x}^{I_1}, \ldots, \sum b_{I_m} \underline{x}^{I_m}\right)$$
$$\underline{x} = (x_1, \ldots, x_n)$$
$$\underline{x}^{I_j} = x_1^{i_{1,j}} \cdots x_n^{i_{n,j}}.$$

An n-dimensional formal group over R is a functor

$$G : \mathcal{C}_R \longrightarrow \mathit{Groups}$$

which is isomorphic to D^n when regarded as a functor to Sets. A formal group *law* is a formal group together with an isomorphism $G \approx D^n$ of set-valued functors.

One source of formal groups is smooth algebraic groups. A smooth algebraic group G over R associates to every R-algebra a group $G(A)$. If A is an adic R-algebra, the assignment

$$\hat{G}(A) = \ker\left\{G(A) \xrightarrow{G(\varepsilon)} G(R)\right\}$$

defines a formal group over R. A choice of local coordinates near the identity gives $\hat{G}$ the structure of a formal group law.

Examples. (1) Let G_a be the additive group. The group $G_a(A)$ is just the abelian group underlying the ring A. The formal group $\hat{G}_a$ associates to every adic R-algebra A the abelian group underlying the augmentation ideal. Taking as a local coordinate of $a \in G_a(A)$ the element a itself, gives $\hat{G}_a$ the structure of a formal group. The associated power series is

$$x \underset{\hat{G}_a}{+} y = x + y.$$

(2) The multiplicative group G_m associates to every ring A its group of units. An element of $\hat{G}_m(A)$ can be written in the form $1 - x$, $x \in \mathfrak{m}$. Take x as a local coordinate. Then the associated formal group law is described by the unique formal power series $x \underset{\hat{G}_m}{+} y$ satisfying

$$(1-x)(1-y) = 1 - (x \underset{\hat{G}_m}{+} y).$$

Thus

$$x \underset{\hat{G}_m}{+} y = x + y - xy.$$

(3) Assume for simplicity that 2 and 3 are invertible in R. After resolving the singularity at ∞, the set of solutions of

$$y^2 = 1 - 2\delta x^2 + \varepsilon x^4$$

in projective 3-space admits a natural group structure, with $x = 0$, $y = 1$ as the identity. Take x as a local coordinate. The formal power series associated to the group structure was computed by Euler:

$$x \underset{F}{+} y = \frac{x\sqrt{R(y)} + y\sqrt{R(x)}}{1 - \varepsilon x^2 y^2}$$
$$R(x) = 1 - 2\delta x^2 + \varepsilon x^4.$$

(4) There is a universal formal group law. This means that there is a ring L and a formal group law F over L with the property that the transformation

$$\mathrm{Hom}_{\mathbf{Rings}}[L, R] \longrightarrow \{\text{Formal group laws over } R\},$$

associating to a homomorphism $L \to R$ the image of F, is a natural equivalence. That such a ring L exists is easy to see. A formal group law over R is described by a formal power series

$$x \underset{F}{+} y = \sum a_{i,j} x^i y^j$$

subject to certain relations amongst the $a_{i,j}$. The ring L is therefore the quotient of the polynomial algebra $\mathbf{Z}[a_{i,j}]$ modulo the ideal generated by these relations. Not so easy to see is the result of Lazard [Laz] that L is free:

$$L \approx \mathbf{Z}[x_1, x_2, \ldots].$$

2 Cohomology theories defined by formal groups

First some results of Quillen [Q]. Let E be a multiplicative cohomology theory. An element $x \in E^2(\mathbf{CP}^\infty)$ whose image under

$$E^2(\mathbf{CP}^\infty) \longrightarrow E^2(S^2) \approx E^0(\mathrm{pt})$$

is a unit is called a *complex orientation* of E. A complex orientation determines a map of ring spectra $MU \to E$, MU being complex cobordism, and gives rise to an isomorphism

$$E^*((\mathbf{CP}^\infty)^n) \approx E^*[\![x_1, \ldots x_n]\!],$$

with

$$x_j = 1 \otimes \cdots \otimes \overset{j}{x} \otimes \cdots \otimes 1.$$

The effect of the tensor product map in E-cohomology is determined by a formal power series

$$F[x_1, x_2] \in E^*(\mathbf{CP}^\infty \times \mathbf{CP}^\infty)$$

which is easily seen to be a formal group law.

The situation is analogous to that of example (4): maps of ring spectra $MU \to E$ are in one to one correspondence with geometrically defined formal group laws over $E^*(\mathrm{pt})$. Quillen [Q] showed that the map

$$L \longrightarrow MU^*,$$

which classifies the formal group law associated with the identity map of MU, is an isomorphism. This raises the possibility that there might be a close correspondence between formal group laws over a ring R and complex oriented cohomology theories with $E^*(\mathrm{pt}) \approx R$. The difficult part of this correspondence is to go from a formal group law to a complex oriented cohomology theory. In good cases the functor

$$E^*(X) = MU^*(X) \underset{MU_*}{\otimes} R \tag{2.1}$$

will satisfy the exactness criteria of Brown's representability theorem, and will therefore be the required cohomology theory. This happens more often that one might expect: the failure of E^* to be exact is supported at those prime ideals of MU_* which are invariant under the cohomology operations. There results a simple criterion for (2.1) to be represented by a cohomology theory. This is known as the Landweber exact functor theorem, and is easiest to state after localizing at a prime p.

After localizing at a prime p, the spectrum MU splits as a ring spectrum into a product of copies of the Brown-Peterson spectrum BP. The Brown-Peterson homology of a point is the polynomial algebra $\mathbb{Z}_{(p)}[\![v_1, v_2, \ldots]\!]$, with $|v_n| = 2p^n - 2$. There is thus a map

$$BP_*(\mathrm{pt}) = \mathbb{Z}_{(p)}[\![v_1, v_2, \ldots]\!] \longrightarrow MU_{*(p)} \approx L_{(p)},$$

and so a natural sequence

$$(p, v_1, v_2, \ldots) \tag{2.2}$$

of elements of $L_{(p)}$. The sequence (2.2) can also be constructed algebraically, at the cost of introducing more of the theory of formal groups.

Exact Functor Theorem [L2]. *Let $f : MU_* \to R$ classify a formal group law F. The functor*

$$E^*(X) = MU^*(X) \underset{MU_*}{\otimes} R$$

is a cohomology theory if and only if the sequence $(f(p), f(v_1), f(v_2), \ldots)$ is regular for every prime p.

Examples. (1) Let R be the graded ring which is $\mathbb{Q}$ in degree 0 and 0 everywhere else, and let $f : MU_* \to R$ classify the additive formal group. For each prime p, $f(p)$ is a unit, so the sequence

$$(f(p), f(v_1), f(v_2), \ldots)$$

is regular. The resulting cohomology theory is ordinary cohomology with rational coefficients.

(2) Let R be the graded ring of Laurent series $\mathbb{Z}[u, u^{-1}]$, $|u| = 2$. Let $MU_* \to R$ classify the twisted multiplicative group

$$x \underset{F}{+} y = x + y + uxy.$$

For each prime p, the element $f(v_1)$ is a unit. The resulting cohomology theory is K-theory.

(3) Let R be the graded ring

$$\mathbf{Z}[\tfrac{1}{2}][\delta, \varepsilon, \Delta^{-1}], \qquad |\delta| = 4, |\varepsilon| = 8,$$

where $\Delta = \varepsilon(\delta^2 - \varepsilon)^2$ is the discriminant of

$$y^2 = 1 - 2\delta x^2 + \varepsilon x^4. \tag{2.3}$$

Let $f : MU_* \to R$ classify the formal group law associated to the elliptic curve (2.3). It can be shown [LRS] that for each prime p, the element $f(v_1)$ is not a zero divisor mod p, and that the element $f(v_2)$ is a unit. The resulting cohomology theory is *Elliptic cohomology*, *Ell*.

(4) Let R be the ring $\mathbf{Z}_{(p)}[v_1, \ldots, v_{n-1}][v_n, v_n^{-1}]$, and let f be the composite

$$MU_* \longrightarrow BP_* \approx \mathbf{Z}_{(p)}[v_1, v_2, \ldots] \longrightarrow R.$$

The map f clearly satisfies the conditions of the exact functor theorem. The resulting cohomology theory goes by the name of $E(n)$ [JW].

3 Characters and generalized cohomology theories

Now to describe $E^*(BG)$ when G is a finite group and when E is a cohomology theory constructed using the exact functor theorem. The results of this section are taken from [HKR].

The motivation for this description is Atiyah's isomorphism [A]

$$K^*(BG) \approx \widehat{R(G)},$$

of the K-theory of BG with the completion of the representation ring of G at its augmentation ideal. The representation ring is most easily studied in terms of characters, so one might hope for a description of the K-theory of BG in terms of class functions:

$$\begin{array}{ccc}
R(G) & \longrightarrow & \widehat{R(G)} \approx K(BG) \\
\Big\downarrow \text{character} & & \Big\downarrow \exists? \\
\text{Class functions: } G \to \mathbf{C} & = & \text{Class functions: } G \to \mathbf{C}.
\end{array}$$

Unfortunately the completion map

$$R(G) \longrightarrow \widehat{R(G)}$$

can have a kernel, so the numinous arrow in the above diagram doesn't exist. The way around this problem is to replace everything by a p-adic analogue for some prime p. Thus instead of K-theory, we use p-adic K-theory, K_p. The field $\mathbf{C}$ gets replaced by its p-adic analogue $\mathbf{C}_p$, the completion of the algebraic closure of $\mathbf{Q}_p$, and instead of class functions on all of G, we consider class functions on the subset $G(1) \subset G$ of elements of order a power of p. With these changes, the completion in $K_p^*(BG)$ is just p-adic completion, and there results a commutative diagram

$$\begin{array}{ccc} R(G) & \xrightarrow{\approx} & K_p^*(BG) \\ \big\downarrow{\scriptstyle \text{character}} & & \big\downarrow \\ \text{Class functions: } G(1) \to \mathbf{C}_p & = & \text{Class functions: } G(1) \to \mathbf{C}_p. \end{array}$$

The new character map extends to an isomorphism

$$\mathbf{C}_p \otimes K_p^*(BG) \approx \mathbf{C}_p\text{-valued class functions on } G(1).$$

How can this character map be described in terms of the K_p-functor alone, without reference to representation theory? Given an element g of $G(1)$ we need to construct a ring homomorphism

$$K_p^*(BG) \longrightarrow \mathbf{C}_p \tag{3.1}$$

which corresponds to evaluation of characters at g. The element g generates a cyclic p-subgroup

$$\mathbf{Z}/|g| \approx \mathbf{Z}/p^n \subset G,$$

and by functoriality, the map (3.1) factors as

$$K_p^*(BG) \longrightarrow K_p^*(B\,\mathbf{Z}/|g|) \longrightarrow \mathbf{C}_p.$$

Since the group $\mathbf{Z}/|g|$ has a preferred generator, the ring $K_p^*(B\,\mathbf{Z}/|g|)$ is canonically isomorphic to $\mathbf{Z}_p[x]/((x+1)^{p^n}-1)$. The map (3.1) is therefore determined by selecting an element $\rho_n \in \mathbf{C}_p$ satisfying

$$(\rho_n + 1)^{p^n} = 1.$$

The element ρ_n must therefore be $e^{2k\pi i/p^n} - 1$ for some k. Naturality of the character map requires that these choices of roots of unity be compatible as n varies:

$$(\rho_n + 1) = (\rho_m + 1)^{p^{m-n}} \qquad \text{if } m > n.$$

The elements ρ_n therefore amount to a homomorphism

$$\mathbf{Q}_p/\mathbf{Z}_p \longrightarrow \text{the } p\text{-Sylow subgroup of the roots of unity}$$

which needs to be an isomorphism if the character map is to be injective. The choice $\rho_n = e^{2\pi i/p^n} - 1$ gives rise to the map described above.

Now to generalize this. Suppose that we are given a ring R, which for simplicity we assume to be a complete Noetherian local domain, with maximal ideal $\mathfrak{m}$, and residue field of characteristic $p > 0$. Suppose also that we are given a formal group law F over R with the property that the classifying map

$$f : MU_* \longrightarrow R$$

satisfies the conditions of the exact functor theorem. Let E be the resulting cohomology theory. By the assumptions on R, there is an integer $n < \infty$ with the property that $f(v_n)$ is a unit, and $f(v_i) \in \mathfrak{m}$ if $i < n$. This integer is the *height* of the mod $\mathfrak{m}$ reduction of F, and will also be referred to as the *height* of E.

Choose a continuous ring homomorphism

$$E \longrightarrow \mathcal{O}_p \subset \mathbf{C}_p,$$

from E to the valuation ring $\mathcal{O}_p$ of $\mathbf{C}_p$. This defines a formal group law (still to be called F) over $\mathcal{O}_p$. Given two elements x and y in the valuation ideal $\mathfrak{p}_p$ of $\mathbf{C}_p$, the formal power series $x \underset{F}{+} y$ will converge. The valuation ideal $\mathfrak{p}_p$, with this new group structure will be denoted $\mathfrak{p}_p^F$. It is known as the *group of points* of F.

3.2 **Theorem** [LT]. *The torsion subgroup, $(\mathfrak{p}_p^F)_{\text{tors}}$, of $\mathfrak{p}_p^F$ is isomorphic to $(\mathbf{Q}_p/\mathbf{Z}_p)^n$.*

Fix an *exponential isomorphism*

$$e : (\mathbf{Q}_p/\mathbf{Z}_p)^n \approx (\mathfrak{p}_p^F)_{\text{tors}}. \tag{3.3}$$

Let $G(n)$ be the set of continuous maps

$$\mathbf{Z}_p^n \longrightarrow G.$$

The set $G(n)$ can be naturally identified with the set

$$G(n) = \{(g_1, \dots, g_n) \mid [g_i, g_j] = 1, g_i^{p^N} = 1 \text{ for } N >> 0\}.$$

The group G acts on $G(n)$ by conjugation:

$$g \cdot (g_1, \dots, g_n) = (gg_1g^{-1}, \dots, gg_ng^{-1}).$$

We are going to construct a character map

$$\chi : E^*(BG) \longrightarrow \text{Hom}_G[G(n), \mathbf{C}_p]$$

from $E^*(BG)$ to the ring of $\mathbf{C}_p$-valued class functions on $G(n)$. Given an element $f \in G(n)$ we need to construct the ring homomorphism

$$E^*(BG) \longrightarrow \mathbf{C}_p$$

corresponding to evaluation at f.

First suppose that G is $(\mathbf{Z}/(p^m))^n$, and that f is reduction mod (p^m). In this case, the cohomology $E^*(BG)$ has been computed [L1].

3.4 Theorem. *The ring $E^*(B(\mathbf{Z}/(p^m))^n)$ is*

$$R[\![x_1]\!]/([p^m](x_1)) \mathop{\widehat{\otimes}}_R \cdots \mathop{\widehat{\otimes}}_R R[\![x_n]\!]/([p^m](x_n))$$
$$= \varprojlim_k R/(\mathfrak{m})^k[\![x_1]\!]/([p^m](x_1)) \otimes_{R/(\mathfrak{m})^k} \cdots \otimes_{R/(\mathfrak{m})^k} R/(\mathfrak{m})^k[\![x_n]\!]/([p^m](x_n)),$$

where x_j is the first Chern class of the character

$$(m_1, \ldots, m_n) \mapsto e^{2\pi i m_j/p^m},$$

and

$$[p^m](x) = x \mathop{+}_F \overset{p^m}{\cdots} \mathop{+}_F x.$$

Evaluation of the characters at f is the map

$$\epsilon : E^*(B(\mathbf{Z}/(p^m))^n) \longrightarrow \mathbf{C}_p$$

defined by

$$\epsilon(x_j) = e(0, \ldots, \overset{j}{\tfrac{1}{p^m}}, \ldots, 0).$$

Now suppose that G is arbitrary. The map $f \in G(n)$ factors as

$$\mathbf{Z}_p^n \longrightarrow (\mathbf{Z}/(p^m))^n \longrightarrow G,$$

for some m. Given $z \in E^*BG$ we define $\chi(z)(f)$ to be the image of z under

$$E^*(BG) \longrightarrow E^*(B(\mathbf{Z}/(p^m))^n) \xrightarrow{\epsilon} \mathbf{C}_p.$$

The main result of [HKR] is

3.5 Theorem. *The generalized character map*

$$E^*(BG) \longrightarrow \mathrm{Hom}_G[G(n), \mathbf{C}_p]$$

extends to an isomorphism

$$\mathbf{C}_p \otimes_R E^*(BG) \approx \mathrm{Hom}_G[G(n), \mathbf{C}_p]. \qquad (3.6)$$

Remark. It isn't really necessary to use $\mathbf{C}_p$ in 3.5; it just makes for the simplest statement. If $g^{p^n} = 1$ for all $g \in G(1)$, it suffices to adjoin the points of order p^n to R and invert $|G|$. More precisely, by the Weierstrass preparation theorem, the formal power series

$$[p^n](x) = x \underset{F}{+} \overset{p^m}{\cdots} \underset{F}{+} x$$

factors uniquely over R into

$$\alpha[x] \cdot u[x] \tag{3.7}$$

where $\alpha[x]$ is a monic polynomial (of degree p^{mn}), and $u(x)$ is a unit in $R[\![x]\!]$. Let $R[p^m]$ be the integral closure of R in the splitting field of α. Then the formulation and conclusion of 3.5 go through with $|G|^{-1}R[p^m]$ playing the role of $\mathbf{C}_p$. This is the smallest ring extension of R which will work, and is often the most natural.

One word about the exponential isomorphism (3.3). In a truly geometric theory, the ring R will not be complete. To apply Theorem 3.5 requires completing the cohomology theory E at a maximal ideal $\mathfrak{m}$. This should be thought of as an artifice of the computation. A good geometric cohomology theory E will have an associated equivariant theory, E_G, with the property that the isomorphism (3.6) is the 'completion' of an isomorphism

$$\begin{aligned} \mathbf{C} \underset{E_*}{\otimes} E_G^*(\mathrm{pt}) &\approx \mathrm{Hom}_G[G_n, \mathbf{C}] \\ G_n &= \mathrm{Hom}[\mathbf{Z}^n, G]. \end{aligned} \tag{3.8}$$

If this is to have a chance of being true, the factorization (3.7) and the the exponential isomorphism need to be independent of the maximal ideal $\mathfrak{m}$.

Over a $\mathbf{Q}$-algebra all formal group laws are isomorphic to the additive formal group. This means that there exists a unique formal power series, the exponential series

$$\exp_F[x] = x + \sum b_n x^{n+1},$$

with the property that

$$\exp_F[x+y] = \exp_F[x] \underset{F}{+} \exp_F[y].$$

The inverse of the exponential series is the logarithm, $\log_F[x]$. The series $\log_F$ can be computed by integrating the invariant differential

$$\log_F[x] = \int_0^x \frac{ds}{F_2[0,s]},$$

where

$$F_2[x,y] = \frac{\partial}{\partial y} F[x,y].$$

Sometimes the exponential series can be used to uniformize the group $\mathfrak{p}_p^F$. This is the case with the multiplicative formal group:

$$\begin{aligned} G_m[x,y] &= x + y - xy, \\ \log_{G_m}[x] &= \int_0^x \frac{ds}{1-s} \\ &= -\ln[1-x], \\ \exp_{G_m}[x] &= 1 - e^{-x}. \end{aligned}$$

In the next section we will see that this is also the case with elliptic cohomology.

4 The case of elliptic cohomology

Recall that the formal group law F of elliptic cohomology is

$$x \underset{F}{+} y = \frac{x\sqrt{R(y)} + y\sqrt{R(x)}}{1 - \varepsilon x^2 y^2}$$

where

$$R(x) = 1 - 2\delta x^2 + \varepsilon x^4.$$

Igusa [I] has shown that for an odd integer n, the series $[n]_F(x)$ is a rational function

$$R_n(x) = \frac{f_n(x)}{g_n(x)} \qquad f(x), g(x) \in \mathbf{Z}[\tfrac{1}{2}][\delta, \varepsilon, \Delta^{-1}][x]$$

with

$$\begin{aligned} f_n(x) &= nx + \ldots \pm \varepsilon^{\frac{n^2-1}{4}} x^{n^2} \\ g_n(0) &= 1. \end{aligned}$$

For n odd, let $R[n]$ be the integral extension of $\mathbf{Z}[\frac{1}{2}][\delta, \varepsilon, \Delta^{-1}]$ obtained by adjoining the roots of f_n. Let $R[\infty]$ be the union of the $R[n]$. The ring $\mathbf{Q} \otimes R[\infty]$ is the natural range of the character map of elliptic cohomology. To set up the character map requires completing at a maximal ideal $\mathfrak{m}$ and choosing an exponential isomorphism.

The formal group law gives the collection of roots of the f_n the structure of a group. As an abstract group it is isomorphic to

$$\mathbf{Q}/\mathbf{Z}[\tfrac{1}{2}] \times \mathbf{Q}/\mathbf{Z}[\tfrac{1}{2}].$$

To specify a particular isomorphism requires a more concrete representation of the rings $R[n]$. This is provided by the theory of automorphic forms.

Over the field of complex numbers, the space of elliptic curves is isomorphic to the space of homothety classes of lattices

$$X = \{\Lambda \subset \mathbf{C}\}/\Lambda \sim z\Lambda, \qquad z \in \mathbf{C}^\times.$$

The Jacobi quartic curve

$$y^2 = 1 - 2\delta x^2 + \varepsilon x^4,$$

is more than just an elliptic curve. It has a natural symmetry

$$(x, y) \mapsto (-x, -y),$$

which corresponds to translation by a preferred point of order 2, $(0, -1)$. This suggests looking at the space

$$E_0 = \{ (\Lambda, \eta) \mid \eta \in \tfrac{1}{2}\Lambda, \quad \eta \notin \Lambda \}$$

of lattices together with a division point of order 2. The group $\mathbf{C}^\times$ acts on E_0 and the orbit space $X_0(2)$ parameterizes the the space of elliptic curves (over $\mathbf{C}$) together with a non-trivial point of order two.

4.2 **Proposition.** *Given $(\Lambda, \eta) \in E_0$, there is a unique meromorphic function*

$$s(z) = s(z; \Lambda, \eta),$$

with the following properties:

$$s(z + \lambda) = s(z) \qquad \text{for } \lambda \in \Lambda \tag{1}$$

$$s(z + \eta) = -s(z);$$

$$s(z) \neq 0 \qquad \text{if } z \notin \Lambda \cup (\Lambda + \eta); \tag{2}$$

$$s(z) = z + O(1) \qquad \text{as } z \to 0. \tag{3}$$

Proof. See for example [Z]. ∎

Since $-s(-z)$ satisfies (1)–(3), s is an odd function.
Let $\psi(z) = 1/s(z)$. Then

$$\psi(z + \lambda) = \psi(z) \qquad \text{for } \lambda \in \Lambda$$

$$\psi(z + \eta) = -\psi(z);$$

$$\psi(z) < \infty \qquad \text{if } z \notin \Lambda \cup (\Lambda + \eta);$$

$$\psi(z) = \frac{1}{z} + O(z) \qquad \text{as } z \to 0.$$

The functions ψ^2 and ψ'^2 are both even and periodic with respect to the lattice

$$\Lambda \cup (\Lambda + \eta). \tag{4.3}$$

Both functions have poles only at the lattice points.

Since

$$\psi(z)^2 = \frac{1}{z^2} + O(1) \qquad \text{as } z \to 0$$
$$\psi'(z)^2 = \frac{1}{z^4} + O(z^{-2}) \qquad \text{as } z \to 0$$

there is a relation

$$\psi'^2 = \psi^4 - 2\delta\psi^2 + \varepsilon$$

for some $\delta = \delta(\Lambda, \eta)$ and $\varepsilon = \varepsilon(\Lambda, \eta)$. It follows that

$$s' = 1 - 2\delta s^2 + \varepsilon s^4$$

and that $s(z)$ is the inverse of the elliptic integral

$$\log_F(w) = \int_0^w \frac{dx}{\sqrt{1 - 2\delta x^2 + \varepsilon x^4}}.$$

The function $s(z)$ is therefore the exponential of the formal group law associated to elliptic cohomology.

The mapping $z \mapsto (s(z), s'(z))$ is the parameterization of the elliptic curve

$$y^2 = 1 - 2\delta x^2 + \varepsilon x^4$$

by $\mathbf{C}/\Lambda$. In this way the elements δ and ε can be regarded as holomorphic functions on the space E_0.

If E is a complex manifold with a $\mathbf{C}^\times$-action, let $\mathcal{O}_E^{2k}$ be the set of holomorphic functions $f : E \to \mathbf{C}$ with the property

$$f(tz) = t^{-k} f(z) \qquad z \in E, t \in \mathbf{C}^\times.$$

The collection of the $\mathcal{O}_E^{2k}$ forms a graded subring $\mathcal{O}_E^* \subset \mathcal{O}_E$ of the holomorphic functions on E.

The function $ts(z; \Lambda, \eta) - s(tz; t\Lambda, t\eta)$ has no zeros and vanishes at the origin. It follows that

$$s(tz; t\Lambda, t\eta) = ts(z; \Lambda, \eta),$$

and hence that

$$s'(tz; t\Lambda, t\eta) = s'(z; \Lambda, \eta)$$
$$\delta(t\Lambda, t\eta) = t^{-2}\delta(\Lambda, \eta)$$
$$\varepsilon(t\Lambda, t\eta) = t^{-4}\varepsilon(\Lambda, \eta).$$

The map

$$\mathbf{Z}[\tfrac{1}{2}][\delta, \varepsilon, \Delta^{-1}] \longrightarrow \mathcal{O}_{E_0}$$

therefore factors through a map of graded rings

$$\mathbf{Z}[\tfrac{1}{2}][\delta, \varepsilon, \Delta^{-1}] \longrightarrow \mathcal{O}^*_{E_0}.$$

Now consider the space of quadruples

$$E_0(n) = \{(\omega_1, \omega_2; \Lambda, \eta)\}$$

consisting of a point $(\Lambda, \eta) \in E_0(2)$, and a pair of points

$$\omega_1, \omega_2 \in \mathbf{C}, \qquad \Im(\frac{\omega_1}{\omega_2}) > 0$$

whose image in $\mathbf{C}/\Lambda$ is a basis of the points of order n. The space $E_0(n)$ admits a $\mathbf{C}^\times$-action

$$t(\omega_1, \omega_2; \Lambda, \eta) = (t\omega_1, t\omega_2; t\Lambda, t\eta),$$

and is a covering space of E_0 with covering group $SL_2(\mathbf{Z}/n)$. The covering map is equivariant. The graded ring $\mathcal{O}^*_{E_0(n)}$ is a Galois extension of $\mathcal{O}^*_{E_0}$ with Galois group $SL_2(\mathbf{Z}/n)$.

For each element $(p, q) \neq 0 \in (\frac{1}{n}\mathbf{Z}/\mathbf{Z})^2$, the ring $\mathcal{O}^*_{E_0(n)}$ contains the function

$$s(a, b) = s(an\omega_1 + bn\omega_2; \Lambda, \eta).$$

Since

$$0 = s(na, nb) = [n](s(a, b)) = \frac{f_n(s(a, b))}{g_n(s(a, b))},$$

the n^2 functions $s(a, b)$ are the roots of f_n. It follows [I] that

$$f_n(x) = \varepsilon^{\frac{n^2-1}{4}} \prod_{(a,b)\in(\frac{1}{n}\mathbf{Z}/\mathbf{Z})^2} (x - s(a, b)).$$

Regard $\mathbf{Q}/\mathbf{Z}[\frac{1}{2}]$ as the subset

$$\left\{ \frac{c}{d} \in \mathbf{Q}/\mathbf{Z} \mid (c, d) = 1,\ d \text{ odd} \right\}.$$

The exponential isomorphism for elliptic cohomology can then be defined by

$$e(a, b) = s(a, b). \tag{4.4}$$

Let K be the field of fractions of $R[\infty]$. Presumably there is an equivariant elliptic cohomology theory, with the property that

$$K \underset{Ell_*}{\otimes} Ell^*_G(\mathrm{pt})$$

is isomorphic to the ring of K-valued class functions on the set of pairs of commuting elements of G. Lacking this, we have to complete the cohomology theory Ell at a maximal ideal, and settle for $\widehat{Ell^*}(BG)$. The maximal ideals of Ell_* correspond to Jacobi quartic curves over finite fields, and the best approximation to Ell_G occurs when the curve is *supersingular* [Silv].

Let $R[p^n]$ be the ring obtained from $\mathbf{Z}[\frac{1}{2}][\delta, \varepsilon, \Delta^{-1}]$ by adjoining the elements

$$s(a,b), \qquad p^n a,\ p^n b \in \mathbf{Z},$$

and let K be the associated field of fractions. Choose a maximal ideal $\mathfrak{m}$ of $\mathbf{Z}[\frac{1}{2}][\delta, \varepsilon, \Delta^{-1}]$ corresponding to a supersingular elliptic curve over a finite field k. Let $\mathfrak{p}$ be the unique extension of $\mathfrak{m}$ to $\widehat{R[p^n]}_{\mathfrak{m}}$. The elements of $\mathfrak{p}$ form a group $\mathfrak{p}^F_p$. The mapping

$$\begin{aligned} e : (\mathbf{Q}_p/\mathbf{Z}_p)^2 &\to (\mathfrak{p}^F_p)_{\mathrm{tors}} \\ (a,b) &\mapsto s(a,b) \end{aligned}$$

is an isomorphism.

4.5 Proposition. *With the above choices, the character map*

$$K \underset{Ell_*}{\otimes} Ell^*(BG) \longrightarrow \mathrm{Hom}_G[G(2), K]$$

is an isomorphism.

5 Concluding remarks

For a conjugacy class $c \subset G$ let Z_c be the centralizer of any of its representatives. The abstract group Z_c is independent of the choice of representative.

The free loop space

$$\mathrm{Map}[S^1, BG]$$

is homotopy equivalent to the disjoint union

$$\coprod_{\substack{\text{conjugacy classes} \\ c \subset G}} BZ_c.$$

The isomorphism (3.6) can be interpreted as showing that modulo torsion, and up to specialization of the cohomology of a point, the ring

$$E_n^*(BG)$$

is isomorphic to

$$E_{n-1}^*(\mathrm{Map}[S^1, BG]).$$

This suggests that to understand the deformations of the genus f_n along a space X, it suffices to understand the deformations of f_{n-1} on the loop space $\mathrm{Map}[S^1, X]$. It also suggests that the geometry and analysis giving rise to elements of $E_{n-1}^*(X)$, when applied to $\mathrm{Map}[S^1, X]$, will give rise to elements of $E_n^*(X)$.

This intuition is upheld in the known examples. Ordinary cohomology is described by integration of forms. Families of Fredholm operators (index bundles), give rise to elements of K-theory. These index bundles can be computed using integration of forms on the loop space. Witten has shown that the elliptic genus of a manifold can be computed as the index of a Dirac-like operator on the loop space, or as an integral on the loop space of the loop space. This last formula gives an intrinsic explanation of the fact that the index of the Dirac-like operator is a modular function.

One trouble with trying to make geometry on loop spaces into a cohomology theory is the fact that loop space functor doesn't patch

$$\mathrm{Map}[S^1, \bigcup U_i] \neq \bigcup \mathrm{Map}[S^1, U_i].$$

This means that geometry and analysis on the loop space of X doesn't sheafify along X, and so some restriction on the geometry to be studied on the loop space needs to be made in order that the Mayer-Vietoris axiom be satisfied.

In the case of the elliptic genus, the operators and the integrands are 'local' in the sense that their values at loops which are amalgams of small loops depend only on their values at the small loops. This goes some distance toward 'sheafifying' analysis on the loop space. Graeme Segal [S] has incorporated this locality principle in his notion of an elliptic object. This seems to get very close to geometrically representing elements of elliptic cohomology.

References

[A] M.F. Atiyah: Characters and cohomology of finite groups. *Inst. Hautes Études Sci. Publ. Math.* 9 (1961), 23–64.

[C] G.E. Carlsson: Equivariant stable homotopy and Segal's Burnside ring conjecture. *Ann. of Math.* 120 (1984), 189–224.

[Ca] H. Cartan: Théories cohomologiques. *Invent. Math.* 35 (1976), 261–271.

[DHS] E. Devinatz, M.J. Hopkins, and J.H. Smith: Nilpotence and stable homotopy theory. I. *Ann. of Math.* 128 (1988), 207–241.

[H] F. Hirzebruch: Topological Methods in Algebraic Geometry, 3rd edition, Springer, Heidelberg, 1966.

[Ho] M.J. Hopkins: Global methods in homotopy theory. *Homotopy Theory, pp. 73–96, Proc. Durham Symposium 1985, Cambridge University Press, Cambridge*, 1987.

[HKR] M.J. Hopkins, N.J. Kuhn, and D.C. Ravenel: Generalized group characters and complex oriented cohomology theories. Preprint.

[I] J.Igusa: On the transformation theory of elliptic functions. *Amer. J. Math.* 81 (1959), 436–452.

[JW] D.C. Johnson and W.S. Wilson: Projective dimension and Brown-Peterson homology. *Topology* 12 (1973), 327–353.

[L1] P.S. Landweber: Complex cobordism of classifying spaces. *Proc. Amer. Math. Soc.* 27 (1971), 175–179.

[L2] P.S. Landweber: Exactness properties of comodules over MU_*MU and BP_*BP. *Amer. J. Math.* 98 (1976), 591–610.

[LRS] P.S. Landweber, D.C. Ravenel, and R.E. Stong: Periodic cohomology theories defined by elliptic curves. To appear.

[Laz] M. Lazard: Sur les groupes de Lie formels à un paramètre. *Bull. Soc. Math. France* 83 (1955), 251–274.

[LT] J. Lubin and J. Tate: Formal complex multiplication in local fields. *Ann. of Math.* 81 (1965), 296–302.

[M] J. Morava: Unpublished research proposal.

[O] S. Ochanine: Sur les genres multiplicatifs définis par des intégrales elliptiques. *Topology* 26 (1987), 143–151.

[Q] D.G. Quillen: The formal group laws of complex and unoriented cobordism theories. *Bull. Amer. Math. Soc.* 75 (1969), 1293–1298.

[R] D.C. Ravenel: Localization with respect to certain periodic homology theories. *Amer. J. Math.* 106 (1984), 351–414.

[S] G. Segal: Elliptic cohomology. *Séminaire Bourbaki 695. Astérisque* 161/162 (1988), 187–201.

[Silv] J.H. Silverman: The Arithmetic of Elliptic Curves, Graduate Texts in Math. 106, Springer-Verlag, Berlin, 1986.

[W] E. Witten: Elliptic genera and quantum field theory. *Comm. Math. Phys.* 109 (1987), 525–536.

[Z] D. Zagier: A note on the Landweber-Stong elliptic genus. *Elliptic Curves and Modular Forms in Algebraic Topology, Lecture Notes in Math. 1326, Springer*, 1988.

Self maps of H-spaces

J. R. Hubbuck

1 Introduction

If $(X, *, \mu)$ is an H-space and K is a based CW-complex, Ioan James [4] proved that the set of based homotopy classes $[K, X]$ is an algebraic loop: the multiplication μ induces an addition '$+$' such that for any $f, g \in [K, X]$, the equations $f + x = g$ and $y + f = g$ have unique solutions for x and y and also there is a unique two sided zero, representing the constant map. If in addition $K = X$, the composition of maps induces a multiplication satisfying a one sided distributive law, $(f + g)h = fh + gh$. When μ is homotopy associative, $[K, X]$ becomes a group and $[X, X]$ a near-ring. So for any H-space X, we describe $[X, X]$ as being a 'near-algebra'. (Such a structure is sometimes called a 'near-loop', but this terminology would be confusing here.)

Let p be a fixed prime and let

$$\alpha : [X, X] \longrightarrow \prod_{k \geq 0} \mathrm{End}\{\bar{\mathrm{H}}^k(X, \mathbf{Z}_p)\}$$

be the obvious homomorphism of near-algebras. So $\alpha(f) = f^{**}$ and the addition on the right hand side is again defined using μ:

$$\text{if} \quad x \in \bar{\mathrm{H}}^k(X, \mathbf{Z}_p) \quad \text{and} \quad \mu^*(x) = x \otimes 1 + 1 \otimes x + \sum x' \otimes x'',$$

$$(f^{**} + g^{**})(x) = f^*(x) + g^*(x) + \sum f^*(x')g^*(x'').$$

We denote the image of α with its near-algebra structure as $\alpha(X)$. In fact $\alpha(X)$ is often a genuine ring.

Examples with $p = 2$:

(1) Let X be a simply connected simple Lie group. Then there are the following ring isomorphisms:

(a) If $X = SU(n), Sp(n), G_2$ or F_4, then $\alpha(X) \cong \mathbf{Z}_2$.

(b) If $X = E_6, E_7$ or E_8, then $\alpha(X) \cong \mathbf{Z}_4$ generated by 1^{**}.

So in both (a) and (b), $\alpha(X)$ is a small commutative ring. (The computations for $Spin(q)$ have not yet been completed.)

(2) Let $X = \Omega Y$ be the space of based loops on a simple Lie group. Then there are the following ring isomorphisms:

(a) If $Y = Sp(n), G_2, F_4, E_7$ or E_8, then $\alpha(X) \cong \mathbf{Z}_2$.

(b) If $Y = E_6$, then $\alpha(X) \cong \mathbf{Z}_8$, generated by 1^{**}.

(c) If $X_q = \Omega\, Spin(q)$, then

$$\begin{aligned} \alpha(X_7) &\cong \mathbf{Z}_2 \times \mathbf{Z}_2, \\ \alpha(X_8) &\cong \mathbf{Z}_2 \times M(2, \mathbf{Z}_2), \\ \alpha(X_{2n+1}) &\cong \mathbf{Z}_2 \quad \text{for} \quad n > 3, \\ \text{and} \quad \alpha(X_{2n}) &\cong \mathbf{Z}_2[x]/(x^2) \quad \text{for} \quad n > 4. \end{aligned}$$

Here $M(n, F)$ denotes the ring of $(n \times n)$-matrices with entries in the finite field F. In (c), $\alpha(X)$ can be realized using loop maps.

Again $\alpha(X)$ is a small commutative ring in all cases except when $X = \Omega Spin(8)$. The case omitted from (2) is $X = \Omega SU(n)$. One can prove that if $2^{t-1} < n \leq 2^t$, then as a group $\alpha(X) \cong \mathbf{Z}_{2^t} \oplus D$, where D is a finite abelian group of exponent $2^s \leq 2^t$. For small values of n, $D = \{0\}$ and it seems probable that this is true in general. When $D = \{0\}$, $\alpha(\Omega SU(n)) \cong \mathbf{Z}_{2^t}$ as a ring generated by 1^{**}.

The computations needed to establish the examples in (1) are routine and those needed in (2) can be found in [3,5]; the calculations for $\Omega Spin(q)$ are due to Zafer Mahmud.

Part of the explanation for the simplicity of the results is that (excluding $\Omega Spin(7)$ and $\Omega Spin(8)$), the spaces X are 2-atomic.

2 Atomic H-Spaces

We assume for the remainder of this note that spaces are connected p-complete based CW-complexes of finite type. What follows is essentially a report of conversations with Michael Crabb, one of us talking about stable homotopy theory and the other about H-spaces. We were particularly influenced by writings of Joel Cohen and Clarence Wilkerson of some years ago. Independently and a little earlier Frank Adams and Nick Kuhn were thinking along related lines; we are grateful for several letters and the manuscript [1] which they sent; these treat atomicity in considerable generality.

With Y a space as described above, $[Y, Y]$ is a profinite compact space. An element $f \in [Y, Y]$ is topologically nilpotent if the sequence $\{f^n\}, n \geq 1$, converges to 0, representing the constant map. Then Y is defined to be *atomic* if each $f \in [Y, Y]$ is either topologically nilpotent or an equivalence.

If Y is atomic and $R = [Y, Y]$ has a natural ring structure with Jacobson Radical $\mathrm{Rad}(R)$, then $\mathrm{Rad}(R)$ is the subspace of topologically nilpotent elements and $[Y, Y]/\mathrm{Rad}(R)$ is a finite field, see for example Remark 2.6 of [1]. There is a similar result for a near-algebra $[X, X]$ when X is an atomic H-space; if N' is the subspace of topologically nilpotent elements, then $[X, X]/N'$ has a natural structure as a finite field. This is discussed below where we seek a more general result.

As an approximation to α, we consider β, the composite homomorphism

$$[X, X] \xrightarrow{\alpha} \prod_{k \geq 0} \mathrm{End}\{\bar{\mathrm{H}}^k(X, \mathbf{Z}_p)\} \longrightarrow \prod_{k \geq 0} \mathrm{End}\{\mathrm{Q}\bar{\mathrm{H}}^k(X, \mathbf{Z}_p)\}.$$

where X is an H-space.

Then $\beta(f + g) = \beta(f) + \beta(g)$ where $+$ on the right hand side is now the standard addition of linear transformations and the image of β is a ring.

The key technical lemma of this note is

2.1 Lemma. *$f \in [X, X]$ is topologically nilpotent if and only if*

$$f^* : \mathrm{Q}\bar{\mathrm{H}}^k(X, \mathbf{Z}_p) \longrightarrow \mathrm{Q}\bar{\mathrm{H}}^k(X, \mathbf{Z}_p)$$

is nilpotent for each k.

This result was certainly known to Adams and the proof requires only minor additions to Section 3 of [1]. The 'only if' part is true because $[X, X] \to \mathrm{End}\{\mathrm{Q}\bar{\mathrm{H}}^k(X, \mathbf{Z}_p)\}$ is continuous by definition of the profinite topology on $[X,X]$, where of course $\mathrm{End}\{\mathrm{Q}\bar{\mathrm{H}}^k(X, \mathbf{Z}_p)\}$ has the discrete topology. The 'if' part follows as there exists an idempotent $h \in [X, X]$ lying in the closure of $\{f^n : n \geq 1\}$. As f^* is nilpotent, it follows that $h^* = 0$ on $\mathrm{Q}\bar{\mathrm{H}}^k(X, \mathbf{Z}_p)$ for each k. Now let $1 + l = 0$ in $[X, X]$. Then

$$0 = (1 + l)h = h + lh = h^2 + lh = (h + l)h.$$

But $(h + l)^* = -1^*$ on $\mathrm{Q}\bar{\mathrm{H}}^k(X, \mathbf{Z}_p)$ for each k and therefore $h + l$ is an equivalence and $h = 0$. The definition of the topology on $[X, X]$ implies that f is topologically nilpotent.

If $\prod_{k \geq 0} \mathrm{End}\{\mathrm{Q}\bar{\mathrm{H}}^k(X, \mathbf{Z}_p)\}$ is given the product topology, a more elegant statement of Lemma 2.1 is that f is topologically nilpotent if and only if $\beta(f)$ is topologically nilpotent. It should be noticed that in the proof of Lemma 2.1, the H-structure of X was used only to define an additive inverse for the identity map. So the result holds equally for any suspended space.

Let X be an H-space and let $N \subset [X, X]$ be defined by

$$N = \{f : fg \text{ is topologically nilpotent for all } g \in [X, X]\}.$$

We can now prove,

2.2 Proposition. *$[X,X]/N$ is a ring.*

Here $[X,X]/N$ denotes the set of equivalence classes given by the relation $f \sim g$ if and only if $f = g+n$ for some $n \in N$. It is possibly helpful to derive this ring structure explicitly although it can be inferred from Section 3 below. Elementary linear algebra and Lemma 2.1 imply that $n_1 + n_2 \in N$ whenever $n_1, n_2 \in N$. One now checks that we do indeed have an equivalence relation. For example, we show symmetry as follows: $f \sim g$ implies that $f = g + n$ for some $n \in N$. There is a unique $k \in [X,X]$ with $g = f + k$ and $k^* = -n^*$ on each $Q\bar{H}^k(X, \mathbf{Z}_p)$. It follows that $k \in N$ and $g \sim f$. Other properties can be proved in a similar manner. Addition and multiplication of equivalence classes are defined by $[f] + [g] = [f + g]$ and $[f][g] = [fg]$. Addition is associative and commutative and so $[X,X]/N$ is an Abelian group and distributive laws hold on both sides.

2.3 Corollary. *If X is an atomic H-space, $[X,X]/N$ is a finite field.*

For $[X,X]/N$ is finite by Lemma 2.1 , as each $\mathrm{End}\{Q\bar{H}^k(X, \mathbf{Z}_p)\}$ is finite dimensional and N coincides with the subset of topologically nilpotent elements. So $[X,X]/N$ is a finite division ring and by Wedderburn's Theorem is a finite field.

By adapting an example given in [1], one can show that any finite field of characteristic p can arise in Corollary 2.3.

3 A Wedderburn-Artin Theorem

3.1 Theorem. *The ring $[X,X]/N$ with the quotient topology is topologically isomorphic to a countable product of matrix algebras $\prod_{i\geq 1} M(n_i, F_i)$ with the product topology, where each F_i is a finite field. The matrix product is unique up to the ordering of the factors.*

We consider the homomorphisms

$$[X,X] \xrightarrow{\beta} \prod_{k\geq 0} \mathrm{End}\{Q\bar{H}^k(X, \mathbf{Z}_p)\} \xrightarrow{\pi_n} \prod_{n\geq k\geq 0} \mathrm{End}\{Q\bar{H}^k(X, \mathbf{Z}_p)\}$$

and denote the image of β as R and the image of $\pi_n\beta$ as R_n. Then R_n is a finite ring and $R = \varprojlim R_n$. The ideals $I_n = ker\{\pi_n | R \to R_n\}$ define the topology on R, which is Hausdorff as $\bigcap I_n = \{0\}$ and β is continuous as each $\pi_n\beta$ is continuous.

The Jacobson Radical of R, $\mathrm{Rad}(R) = \{\ x : xy$ is right quasi-regular for all $y \in R\}$ where z is quasi-regular if $(1 + z)w = 1$ for some w, [2]. The ring R_n is finite and so for any $r \in R_n$, there exists $\alpha > 0$ such that r^α is an idempotent. If $r \in \mathrm{Rad}(R_n)$, then $-r^\alpha$ is right quasi-regular and by multiplying $(1 - r^\alpha)w = 1$ on the left by r^α, it follows that $0 = r^\alpha$. So $\mathrm{Rad}(R_n) = \{\ r : rs$ is nilpotent for each $s \in R_n\}$. Now

$x \in R$ is topologically nilpotent if and only if $\pi_n(x)$ is nilpotent in R_n for each n. It follows that $\mathrm{Rad}(R) = \{x : xy \text{ is topologically nilpotent for all } y \in R\}$. Therefore

$$[X,X]/N \quad\cong\quad R/\mathrm{Rad}(R) \quad\cong\quad \varprojlim\{R_n/\mathrm{Rad}(R_n)\}.$$

The Wedderburn-Artin Theorem implies that $R_n/\mathrm{Rad}(R_n) \cong \prod_{1\le i\le s_n} M(n_i, F_i)$ where the F_i are finite fields. Let

$$\tau_{n+1} : R_{n+1}/\mathrm{Rad}(R_{n+1}) \longrightarrow R_n/\mathrm{Rad}(R_n)$$

be the projection. Any surjective homomorphism

$$\prod_{1\le j\le s_{n+1}} M(m_j, F_j) \longrightarrow M(n_i, F_i)$$

must map all but one factor trivially and one factor isomorphically, as the only ideals in $M(m, F)$ are the trivial ideals. So we can write

$$\tau_{n+1} : \prod_{1\le i\le s_{n+1}} M(n_i, F_i) \longrightarrow \prod_{1\le i\le s_n} M(n_i, F_i), \quad s_n \le s_{n+1},$$

where the first s_n factors are mapped isomorphically and the remainder are mapped to zero. Thus

$$R/\mathrm{Rad}(R) \cong \prod_{i\ge 1} M(n_i, F_i).$$

The isomorphism in the theorem is a homeomorphism as $[X,X]/N$ is compact and $R/\mathrm{Rad}(R)$ is Hausdorff, which completes the proof.

The fields F_i which arise in Theorem 3.1 are the fields given by Corollary 2.3 when X is expressed as a product of atomic H-spaces with increasing connectivities. The particular H-multiplication μ chosen for X plays no role in the final result.

Theorem 3.1 gives a little insight into the structure of $\alpha(X)$. It does not explain why the near-ring structures of the examples in the introduction are rings, but when $\alpha(X)$ is a ring, the quotient ring $\alpha(X)/\mathrm{Rad}(\alpha(X))$ is isomorphic to the $R/\mathrm{Rad}(R)$ of Theorem 3.1.

Finally, it is clear that the comments following Lemma 2.1 imply that Theorem 3.1 remains true as stated if the H-space X is replaced by any p-complete based CW-complex of finite type which is a suspension.

References

[1] J.F.Adams and N.J.Kuhn: Atomic Spaces and Spectra. To appear in *Proc. Edinburgh Math. Soc.*

[2] N.J.Divinsky: Rings and Radicals, George Allen and Unwin Ltd, London, 1965.

[3] J.R.Hubbuck: Unstable atomicity and loop spaces on Lie groups. *Math. Z.* 198 (1988), 317–330.

[4] I.M.James: On H-spaces and their homotopy groups. *Quart. J. Math. Oxford* 11 (1960), 161–179.

[5] Z. Mahmud: Atomicity and $\Omega Spin(q)$. To appear in *Proc. Roy. Soc. Edinburgh, Sect. A.*

Character rings in algebraic topology

Nicholas J. Kuhn*

1 Introduction

Let $R(G)$ be the complex representation ring of a finite group G. The assignment

$$G \mapsto R(G)$$

is a Mackey functor, i.e., a functor from groups to rings endowed with induction (transfer) maps. Typically one studies this functor via characters. One lets $C(G)$ be the ring of complex-valued functions on G invariant under conjugation, and then one defines a natural map of rings

$$\chi: R(G) \mapsto C(G)$$

by $\chi(M)(g) = \text{trace}\{g: M \to M\}$. There is a simple formula defining induction between character rings making χ a map of Mackey functors.

Why is χ so important? There are three reasons:

(i) χ is injective,

(ii) χ is as surjective as possible: $R(G) \subset C(G)$ is a maximal $\mathbf{Z}$-lattice, so that $R(G) \otimes_{\mathbf{Z}} \mathbf{C} \simeq C(G)$, and

(iii) $C(G)$ is very concrete and easy to work with.

Analogues of $R(G)$ are well known to topologists: given a multiplicative cohomology theory E^*, the assignment

$$G \mapsto E^*(BG)$$

is a Mackey functor.

In this paper, I discuss some natural generalizations of the classical character ring that can be used to detect $E^*(BG)$, for various E^*, in much the same way that $C(G)$ detects $R(G)$. Two recent projects fit into this picture: my joint work with

*Research partially supported by the National Science Foundation

M. Hopkins and D. Ravenel on v_n-periodic complex oriented theories [HKR], and the reinterpretation and extension of Quillen's work on $H^*(BG)$ by J. Lannes and L. Schwartz and others (see e.g., [HLS]).

To explain how one generalizes $C(G)$, observe that $G \simeq \mathrm{Hom}(\mathbf{Z}, G)$. Letting G act trivially on $\mathbf{C}$ and by conjugation on $\mathrm{Hom}(\mathbf{Z}, G)$, we have a natural isomorphism of $\mathbf{C}$-algebras

$$C(G) \simeq \mathrm{Map}_G(\mathrm{Hom}(\mathbf{Z}, G), \mathbf{C}).$$

We will replace $\mathbf{Z}$ by other groups Γ and $\mathbf{C}$ by other fields F. In the examples, the choice of Γ seems to be vitally linked to the geometry underlying the cohomology theory E^*.

We discuss the organization of the paper.

In Section 2 and Section 3 we discuss formal properties of our generalized character rings. In particular, they are Mackey functors, and one is naturally led to inverse limits over categories of detecting subgroups.

The next two sections have parallel descriptions of the two projects referred to above: Section 4 reviews the $H^*(BG)$ results, while Section 5 describes the characters for complex oriented theories. The Lannes-Schwartz work leads to a purely group theoretic criterion guaranteeing, for a finite p-group G, that $H^*(BG; \mathbf{Z}/p)$ has an infinite-dimensional A-module summand in the kernel of Quillen's map. The criterion is satisfied when G is the quaternionic group of order 8. Reasoning in an analogous manner about the nth Morava K-theory $K(n)^*(BG)$, we give a group theoretic criterion that would imply that $K(n)^*(BG)$ is *not* concentrated in even degrees,as has been conjectured. Fortunately, or unfortunately, we have yet to find a group satisfying our condition — indeed, there are theorems in group theory hinting that it cannot be done.

In Section 6, we briefly describe how our character rings extend to equivariant cohomology theories. As a simple example, we show how classical character theory extends to a computation of $K_G(X) \otimes \mathbf{C}$ in non-equivariant terms.

The last section touches on the possibility of assembling character rings into simplicial objects.

We would like to thank Mike Hopkins for providing the proof of Lemma 5.6 and Leonard Scott for bringing our attention to the theorem of John Thompson used in proving Proposition 5.7.

2 Character rings: definitions and examples

We begin by defining our rings of class functions. Given two topological groups Γ and G, we let $\mathrm{Hom}(\Gamma, G)$ denote the space of continuous homomorphisms. This becomes a (left) G-space by letting G act on itself by conjugation. In the examples we consider, G will be a finite group and $\mathrm{Hom}(\Gamma, G)$ will be finite and discrete. Let F be a field (possibly graded).

2.1 Definition. $C_{\Gamma,\mathsf{F}}(G) = \mathrm{Map}_G(\mathrm{Hom}(\Gamma, G), \mathsf{F})$.

Thus an element of $C_{\Gamma,\mathsf{F}}(G)$ is a function

$$f \colon \mathrm{Hom}(\Gamma, G) \longrightarrow \mathsf{F},$$

constant on G-orbits. $C_{\Gamma,\mathsf{F}}(G)$ is an F-algebra using pointwise multiplication and addition of functions. It is clearly a contravariant functor of G, and a covariant functor of Γ and F.

To show that $C_{\Gamma,\mathsf{F}}(G)$ is a Mackey functor, we need to define induction. The classical formula [S, page 30] generalizes.

2.2 Definition. *Let H be a subgroup of a finite group G. Define*

$$ind_H^G \colon C_{\Gamma,\mathsf{F}}(H) \longrightarrow C_{\Gamma,\mathsf{F}}(G)$$

by the formula

$$ind_H^G(f)(\alpha) = \sum_{gH \in (G/H)^{\mathrm{Im}(\alpha)}} f(g \cdot \alpha),$$

where $f \colon Hom(\Gamma, H) \to \mathsf{F}$, $\alpha \colon \Gamma \to G$, and $g \cdot \alpha$ denotes α composed with conjugation by g.

It is straightforward to check that this is well-defined.

2.3 Proposition. *With this structure, the assignment $G \mapsto C_{\Gamma,\mathsf{F}}(G)$ is a Mackey functor.*

Proof. We need only check that the double coset formula holds. Let K and H be subgroups of G and $\mathrm{res}_G^K \colon C_{\Gamma,\mathsf{F}}(G) \to C_{\Gamma,\mathsf{F}}(K)$ the restriction. Letting K_g denote $K \cap gHg^{-1}$, there is an isomorphism of left K-sets

$$\coprod_{KgH} K/K_g = G/H,$$

given by sending kK_g to kgH. Thus

$$\begin{aligned} \mathrm{res}_G^K(\mathrm{ind}_H^G(f))(\alpha) &= \sum_{gH \in (G/H)^{\mathrm{Im}(\alpha)}} f(g \cdot \alpha) \\ &= \sum_{KgH} \sum_{kK_g \in (K/K_g)^{\mathrm{Im}(\alpha)}} f(kg \cdot \alpha) \\ &= \sum_{KgH} \mathrm{ind}_{K_g}^K(f)(g \cdot \alpha), \end{aligned}$$

as needed. ∎

2.4 Examples. (1) $\Gamma = \mathbf{Z}$. Then $\mathrm{Hom}(\mathbf{Z}, G) = G$, and $C_{\mathbf{Z},\mathbf{C}}(G)$ is the usual character ring detecting $R(G)$.

(2) $\Gamma = \mathbf{Z}[\frac{1}{p}]$. Then $\mathrm{Hom}(\mathbf{Z}[\frac{1}{p}], G) = G_{\mathrm{reg}}$, the "$p$-regular" elements in G of order prime to p. Note that the inclusion $G_{\mathrm{reg}} \subset G$ is induced by $\mathbf{Z} \to \mathbf{Z}[\frac{1}{p}]$. The ring $C_{\mathbf{Z}[\frac{1}{p}],\mathbf{F}}(G)$ arises in Brauer character theory. Let A be a p-adic ring of integers with quotient field K and residue field k. Assume that A contains the $|G|$th roots of 1. Let $R_K(G)$ (respectively $R_k(G)$) be the Grothendieck ring of $K[G]$ ($k[G]$) modules. There is a commutative diagram of ring homomorphisms:

$$\begin{array}{ccc} R_K(G) & \xrightarrow{\chi_K} & C_{\mathbf{Z},K}(G) \\ d\downarrow & & \downarrow \\ R_k(G) & \xrightarrow{\chi_k} & C_{\mathbf{Z}[\frac{1}{p}],K}(G) \end{array}$$

where χ_K is the usual character map, χ_k is the Brauer character map, and d is the "decomposition" map [S]. Both χ_K and χ_k are inclusions of maximal $\mathbf{Z}$-lattices, and d is surjective.

(3) $\Gamma = \mathbf{Z}_p$, the p-adic integers. $\mathrm{Hom}(\mathbf{Z}_p, G)$ is the set of "p-unipotent" elements in G, the elements of order a power of p. Where does $C_{\mathbf{Z}_p,\mathbf{F}}(G)$ occur naturally as a character ring? Let $K(BG)_p$ denote the p-adic completion of the complex K-theory of BG. Atiyah's theorem [A] and the methods of [K1] show that there is a unique continuous extension of χ in the diagram

$$\begin{array}{ccc} R(G) & \xrightarrow{\chi} & C_{\mathbf{Z},\overline{\mathbf{Q}}_p}(G) \\ \downarrow & & \downarrow \\ K(BG)_p & \xrightarrow{\chi_p} & C_{\mathbf{Z}_p,\overline{\mathbf{Q}}_p}(G) \end{array}$$

and that χ_p is the embedding of a maximal $\mathbf{Z}_p$-lattice.

(4) $\Gamma = \mathbf{Z}^n$. Then $\mathrm{Hom}(\mathbf{Z}^n, G)$ is the set of n-tuples of commuting elements in G. The case $n = 2$ seems to occur in "elliptic" settings — see e.g., work by S. Norton on "montrous moonshine" [N]. Note that if C is a complex elliptic curve, $\pi_1(C) = \mathbf{Z} \times \mathbf{Z}$. See Section 6 for one Mackey functor detected by $C_{\mathbf{Z}^n,\mathbf{F}}(G)$.

(5) $\Gamma = (\mathbf{Z}_p)^n$. This is the p-adic version of (4). $\mathrm{Hom}(\mathbf{Z}_p^n, G)$ is the set of n-tuples of elements in G generating an abelian p-group. For "naturally occurring" instances of $C_{\mathbf{Z}_p^n,\overline{\mathbf{Q}}_p}(G)$ and $C_{\mathbf{Z}_p^n,\overline{\mathbf{F}}_p}(G)$, see Section 5; these rings try to detect $E^*(BG)$, where E^* is a p-local v_n-periodic oriented theory.

(6) $\Gamma = (\mathbf{Z}/p)^n$. Then $\mathrm{Hom}((\mathbf{Z}/p)^n, G)$ is the set of n-tuples generating an elementary abelian p-group. Quillen's theorem essentially says that, if $n \geq \mathrm{rank}(G)$, then $C_{(\mathbf{Z}/p)^n,\mathbf{Z}/p}(G)$ detects $H^*(BG; \mathbf{Z}/p)$ up to F-isomorphism. See Section 4.

(7) $\Gamma = F_n$, the free group on n generators. As a reminder that Γ might not be abelian, we point out that $G^n = \mathrm{Hom}(F_n, G)$. See Section 7.

3 Detecting families and counting orbits

This section exploits the obvious observation that the image of a homomorphism $\Gamma \to G$ is simultaneously a quotient group of Γ and a subgroup of G. Thus, for example, it is intuitively clear that if Γ is abelian, $\mathrm{Hom}(\Gamma, G)$ should only depend on $\mathrm{Hom}(\Gamma, A)$ where A runs through the abelian subgroups of G.

To be more formal, let $\mathcal{S}(G)$ be the category whose objects are the subgroups of G, and whose morphisms are generated by inclusions $H < K$ and conjugation by elements $g \in G$, $c_g\colon H \to gHg^{-1}$. Let $\mathcal{J}(G) \subset \mathcal{S}(G)$ just have inclusions for morphisms.

3.1 Definition. *Let $\Gamma(G)$ be the full subcategory of $\mathcal{S}(G)$ with objects those subgroups of G occurring as quotient groups of Γ. Let $\mathcal{J}\Gamma(G) = \Gamma(G) \cap \mathcal{J}(G)$.*

The observation at the beginning of this section can be more precisely stated as:

3.2 Proposition. *The natural map*

$$C_{\Gamma,\mathsf{F}}(G) \longrightarrow \lim_{H \in \Gamma(G)} C_{\Gamma,\mathsf{F}}(H)$$

is an isomorphism. The same is true if $\Gamma(G)$ is replaced by any full subcategory of $\mathcal{S}(\mathcal{G})$ containing $\Gamma(G)$.

Proof. There is an evident bijection of G-sets

$$\operatorname*{colim}_{H \in \mathcal{J}\Gamma(G)} \mathrm{Hom}(\Gamma, H) \longrightarrow \mathrm{Hom}(\Gamma, G).$$

Taking G-invariant maps into F yields the isomorphism, noting that, for any contravariant functor F,

$$\lim_{H \in \Gamma(G)} F(H) = \Big(\lim_{H \in \mathcal{J}\Gamma(G)} F(H)\Big)^G.$$

3.3 Examples. (1) Let $\mathcal{A}(G)$ be the full subcategory of $\mathcal{S}(G)$ with the abelian subgroups as objects. Then, for any n,

$$C_{\mathbf{Z}^n,\mathsf{F}}(G) \simeq \lim_{\mathcal{A}(G)} C_{\mathbf{Z}^n,\mathsf{F}}(A).$$

(2) Let $\mathcal{E}_p(G)$ be the full subcategory of $\mathcal{S}(G)$ with the elementary abelian p-groups as objects. Then, if V is an F_p-vector space,

$$C_{V,\mathsf{F}}(G) \simeq \lim_{\mathcal{E}_p(G)} C_{V,\mathsf{F}}(E).$$

(3) Let $\mathcal{C}(G)$ be the full subcategory of $\mathcal{S}(G)$ with the cyclic groups as objects. Then

$$C_{\mathbf{Z},\mathsf{F}}(G) \simeq \lim_{\mathcal{C}(G)} C_{\mathbf{Z},\mathsf{F}}(C).$$

Our little observation can be utilized in another way. Note that there is a natural isomorphism of G-sets

$$\coprod_Q \mathrm{Epi}(\Gamma, Q) \times_{\mathrm{Aut}(Q)} \mathrm{Mono}(Q, G) = \mathrm{Hom}(\Gamma, G),$$

where the union is over isomorphism classes of finite groups. For any fixed G, this is clearly just a finite disjoint union. The group $\mathrm{Aut}(Q)$ acts freely on both $\mathrm{Epi}(\Gamma, Q)$ and $\mathrm{Mono}(Q, G)$. Furthermore, $\mathrm{Map}_G(\mathrm{Mono}(Q, G), \mathsf{F})$ is easily seen to be a quotient Mackey functor of $C_{Q,\mathsf{F}}(G)$. These observations imply

3.4 Proposition.

$$C_{\Gamma,\mathsf{F}}(G) \simeq \prod_Q |\mathrm{Epi}(\Gamma, Q)| \times \frac{1}{|\mathrm{Aut}(Q)|} \times \mathrm{Map}_G(\mathrm{Mono}(Q, G), \mathsf{F}),$$

as Mackey functors. (Here $n \times R$ means the n-fold product $R \times R \times \cdots \times R$.)

To see how this proposition can be used, we generalize counting arguments that occurred in [K2] and [HKR]. Let G be a finite group. The group ring $\mathsf{F}_p[\mathrm{Out}(G)]$ acts on $C_{\Gamma,\mathsf{F}_p}(G)$ as does the larger F_p-algebra $A(G,G) \otimes \mathsf{F}_p$ generated by inductions composed with homomorphisms. If G is a p-group, this latter algebra is known to be isomorphic to the stable endomorphism ring $\{BG_+, BG_+\} \otimes \mathsf{F}_p$ (as in [Ma]). Let e be an idempotent in either $\mathsf{F}_p[\mathrm{Out}(G)]$ or $A(G,G) \otimes \mathsf{F}_p$. Let $f_e(n) = \dim_{\mathsf{F}_p} eC_{\mathsf{Z}_p^n,\mathsf{F}_p}(G)$. In Section 5, it will be shown that this function of n has topological meaning: $f_e(n)$ is the nth Morava K-theory Euler characteristic of the spectrum eBG_+. Note that $f_1(n) = |\mathrm{Hom}(\mathsf{Z}_p^n, G)/G|$.

Our application of Proposition 3.4 is

3.5 Proposition. *For all p-groups G and e, $f_e(n)$ is a polynomial in p^n, with rational coefficients, of degree $\leq d$ where p^d is the order of the largest abelian subgroup of G.*

To prove this, if Q is an abelian p-group, let $f_Q(n) = |\mathrm{Epi}(\mathsf{Z}^n, Q)|$. By Proposition 3.4,

$$f_e(n) = \sum_Q \frac{1}{|\mathrm{Aut}(Q)|} \times \dim_{\mathsf{F}_p} e\mathrm{Map}_G(\mathrm{Mono}(Q, G), \mathsf{F}_p) \times f_Q(n),$$

so that $f_e(n)$ is a linear combination of functions $f_Q(n)$. It suffices to verify the next lemma.

3.6 Lemma. *If $|Q| = p^d$, $f_Q(n)$ is a polynomial in p^n of degree d.*

Proof. Observe that

$$\mathrm{Hom}(\mathsf{Z}_p^n, Q) = \coprod_{Q' < Q} \mathrm{Epi}(\mathsf{Z}_p^n, Q').$$

Noting that $|\mathrm{Hom}(\mathbf{Z}_p^n, Q)| = |Q^n| = p^{nd}$, this implies that

$$f_Q(n) = p^{nd} - \sum_{Q' \lneq Q} f_{Q'}(n).$$

The result follows by induction on $|Q|$.

3.7 **Remarks.** (i) The denominators in the rational coefficients occurring in $f_e(n)$ are just due to the factors $1/|\mathrm{Aut}(Q)|$.

(ii) For related, but slightly different, counting arguments, see [HKR, §3].

(iii) For some examples of $f_e(n)$'s when G is abelian, see [K2, §6].

4 H*(BG) revisited

In this section, $H^*(G)$ denotes $H^*(BG; \mathsf{F}_p)$, and V denotes a finite-dimensional F_p-vector space. Let $\mathcal{U}$ be the category of unstable modules over the Steenrod algebra, and let $\mathcal{K}$ be the category of unstable A-algebras. Lannes and Schwartz [HLS] have re-examined Quillen's work on $H^*(G)$ using unstable A-module technology. We place their results in our character ring context.

There is a natural map

$$\mathsf{F}_p[\mathrm{Hom}(V,G)] \longrightarrow \mathrm{Hom}_{\mathcal{U}}(H^*(G), H^*(V)),$$

where $\mathsf{F}_p[\mathrm{Hom}(V,G)]$ is an F_p-vector space with basis $\mathrm{Hom}(V,G)$. Since inner automorphisms of G induce the identity in cohomology, this factors through

$$\mathsf{F}_p[\mathrm{Hom}(V,G)/G] \longrightarrow \mathrm{Hom}_{\mathcal{U}}(H^*(G), H^*(V)).$$

Taking (profinite) duals yields a natural map

$$\chi_G : \mathrm{Hom}_{\mathcal{U}}(H^*(G), H^*(V))^* \longrightarrow C_{V,\mathsf{F}_p}(G).$$

4.1 **Theorem** [HLS]. *χ_G is an isomorphism of Mackey functors. Furthermore, taking Spec of both rings yields a natural bijection*

$$\mathrm{Hom}(V,G)/G \longrightarrow \mathrm{Hom}_{\mathcal{K}}(H^*(G), H^*(V)).$$

We sketch the proof. Firstly, it is straightforward to check that χ_G commutes with induction maps — using the double coset formula, one is reduced to the following well-known lemma.

4.2 **Lemma.** *If V' is a proper subspace of the F_p-vector space V, the cohomology transfer $\mathrm{ind}_{V'}^{V} : H^*(V') \to H^*(V)$ is the zero map.*

The proof that χ_G is an isomorphism then uses four deep facts:

(1) [Q1] $\alpha_G: H^*(G) \to \lim_{\mathcal{E}_p(G)} H^*(E)$ is an F-isomorphism;

(2) [LS] a map $M \to N$ in $\mathcal{K}$ is an F-isomorphism if and only if the induced map $\mathrm{Hom}_{\mathcal{U}}(N, H^*(V)) \to H^*_{\mathcal{U}}(M, H^*(V))$ is an isomorphism for all V;

(3) [C1,Mi] $H^*(V)$ is injective in $\mathcal{U}$; and

(4) [AGM] χ_E is an isomorphism if E is elementary abelian.

Armed with these, it is easy to show that χ_G is an isomorphism. Consider the diagram

$$\begin{array}{ccc}
\mathrm{Hom}_{\mathcal{U}}(H^*(G), H^*(V))^* & \xrightarrow{\chi_G} & C_{V,\mathbb{F}_p}(G) \\
\downarrow \alpha & & \\
\mathrm{Hom}_{\mathcal{U}}(\lim_{\mathcal{E}_p(G)} H^*(E), H^*(V))^* & & \downarrow \gamma \\
\downarrow \beta & & \\
\lim_{\mathcal{E}_p(G)} \mathrm{Hom}_{\mathcal{U}}(H^*(E), H^*(V))^* & \xrightarrow{\lim \chi_E} & \lim_{\mathcal{E}_p(G)} C_{V,\mathbb{F}_p}(E).
\end{array}$$

The map α is an isomorphism by facts (1) and (2), β is an isomorphism by (3), and (4) implies that $\lim \chi_E$ is an isomorphism. In Example 3.3 (2), we showed that γ is an isomorphism; thus, χ_G is also.

Of course, χ_G is functorial in V as well as G. This is the functoriality exploited by Lannes and Schwartz when they assign an "analytic functor" to an unstable A-module [HLS]. This leads to a rather practical group-theoretic way to prove that $H^*(G)$ has many nilpotent elements.

If S is a simple $\mathbb{F}_p[\mathrm{Out}(G)]$ or $A(G,G) \otimes \mathbb{F}_p$ module, let e_S be an associated idempotent, so that $e_S M = 0$ exactly when S does not occur as a composition factor in M.

4.3 Theorem. *Let G be a p-group of rank d, and let e be an idempotent in either $\mathbb{F}_p[\mathrm{Out}(G)]$ or $A(G,G) \otimes \mathbb{F}_p$. Then*

$$e\, C_{(\mathbb{Z}/p)^d, \mathbb{F}_p}(G) = 0 \iff e H^*(G) \subset \ker(\alpha_G),$$

where $\alpha_G: H^*(G) \to \lim_{\mathcal{E}_p(G)} H^*(E)$ is the natural map. In particular, if S is not a composition factor in $C_{(\mathbb{Z}/p)^d, \mathbb{F}_p}(G)$, then $e_S H^*(G)$ is all nilpotent.

Before proving this we look at some examples.

4.4 Example. Let $G = Q_8$, the quaternion group of order 8. $\mathrm{Out}(Q_8) \simeq \Sigma_3$, permutations of i, j, k. The $\mathrm{Out}(Q_8)$-set $\mathrm{Hom}(\mathbb{Z}/2, Q_8)$ has trivial action, since $Z(Q_8) = \{\pm 1\}$ is the only nontrivial elementary abelian subgroup. However, $\mathbb{F}_2[\mathrm{Out}(Q_8)]$ has *two* simple modules, the trivial module and a two-dimensional one. We conclude that $H^*(Q_8)$ decomposes (over the Steenrod algebra) as $H^*(Q_8) = M_0 \oplus M_1 \oplus M_1$, where $M_1 \subset \ker(\alpha_{Q_8})$. We note that $M_0 \simeq H^*(SL_2(\mathbb{F}_3))$, and $M_1 \simeq H^*(\Sigma^{-1} BS^3/BN)$ [MP].

4.5 Example. If $\alpha_G: H^*(G) \to \lim_{\mathcal{E}_p(G)} H^*(E)$ is monic, it follows that every simple $\mathrm{Out}(G)$ (and $A(G,G)\otimes \mathbf{F}_p$) module occurs as a composition factor in $C_{V,\mathbf{F}_p}(G)$ with $V \geq \operatorname{rank} G$. In [Q2], it is shown that if α_G is monic, then so is $\alpha_{\mathbf{Z}/p \wr G}$. In particular, if G is the p-Sylow subgroup of a symmetric group, then the map α_G is monic.

We sketch the proof of Theorem 4.3. Reasoning as in Section 3 (e.g., Proposition 3.4), shows that

$$e\,C_{(\mathbf{Z}/p)^d,\mathbf{F}_p}(G) = 0 \iff \forall V,\ e\,C_{V,\mathbf{F}_p}(G) = 0.$$

Let $\mathcal{H}^*(G) = \lim_{\mathcal{E}_p(G)} H^*(E)$. In the language of Lannes and Schwartz, $\mathcal{H}^*(G)$ is "nil-closed", and since α_G is an F-isomorphism, $\mathcal{H}^*(G)$ is the nil-closure of $H^*(G)$. Lannes and Schwartz show that a nil-closed object M is *equivalent* to the functor $V \mapsto \operatorname{Hom}_{\mathcal{U}}(M, H^*(V))^*$. Thus $\mathcal{H}^*(G)$ is equivalent to $V \mapsto C_{V,\mathbf{F}_p}(G)$. Thus, if e is an idempotent,

$$\begin{aligned} e\,C_{V,\mathbf{F}_p}(G) = 0 \text{ for all } V &\iff e\mathcal{H}^*(G) = 0 \\ &\iff eH^*(G) \subset \ker \alpha_G. \end{aligned}$$

4.6 Remark. If $e \neq 0$, then $eH^*(G)$ is always infinite dimensional, a consequence of the fact that $\{BG_+, BG_+\}$ is torsion free: if $eH^*(G)$ were finite dimensional, then the order of the identity in $\{eBG_+, eBG_+\}$ would be finite.

5 Characters for complex-oriented theories

We first summarize some theorems that will appear in [HKR]. As motivation, recall that

$$\chi_G: R(G) \longrightarrow C_{\mathbf{Z},\mathbf{C}}(G)$$

is an embedding of a maximal $\mathbf{Z}$-lattice, and that

$$C_{\mathbf{Z},\mathbf{C}}(G) \simeq \lim_{\mathcal{C}(G)} C_{\mathbf{Z},\mathbf{C}}(C).$$

This suggests that perhaps $R(G) \simeq \lim_{\mathcal{C}(G)} R(C)$. This is almost true: Artin's Theorem [S] says that there is an isomorphism

$$R(G) \otimes \mathbf{Z}[1/|G|] \simeq \lim_{\mathcal{C}(G)} R(C) \otimes \mathbf{Z}[1/|G|].$$

It turns out that the splitting principle and equivariant general nonsense suffice to prove an analogous theorem.

5.1 Theorem [HKR]. *Let E^* be any complex-oriented theory. For any finite group G, the natural map*

$$E^*(BG) \otimes \mathbf{Z}[1/|G|] \longrightarrow \lim_{\mathcal{A}(G)} E^*(BA) \otimes \mathbf{Z}[1/|G|]$$

is an isomorphism.

This suggests that character rings for detecting complex-oriented theories should be of the form $C_{\Gamma,\mathsf{F}}(G)$ with Γ abelian, so that $\Gamma(G) \subset \mathcal{A}(G)$, and Proposition 3.2 applies. Our main discovery is that this can be done, with the rank of Γ corresponding to the type of v_n-periodicity present in E^*.

To describe one version of our result, we recall some notation. Localized at a prime p, MU is equivalent to a wedge of suspensions of the Brown-Peterson spectrum BP. The coefficients are $BP_* = \mathbf{Z}_{(p)}[v_1, v_2, \ldots]$ with $|v_i| = 2p^i - 2$. A complex-oriented theory E^* is *v_n-periodic* if v_n is a unit in E^*. For example, K-theory is v_1-periodic. Let $I_n \subset BP_*$ be the ideal $(p, v_1, v_2, \ldots, v_{n-1})$. Finally, recall that to a complex-oriented theory E^*, there is an associated formal group law F, and $[m](x) \in E^*[[x]]$ denotes the m-fold formal sum $x +_F x +_F \cdots +_F x$.

5.2 Theorem [HKR]. *Let E^* be a multiplicative v_n-periodic cohomology theory such that the coefficient ring is a characteristic 0 domain and is complete in the I_n-adic topology. Let $\mathsf{F}(E^*)$ be the graded field of fractions of E^* with solutions to $[p^i](x) = 0$ adjoined, for all i, as elements in degree 2. Then there is an isomorphism of Mackey functors*

$$\chi_G \colon E^*(BG) \otimes_{E^*} \mathsf{F}(E^*) \simeq C_{\mathbf{Z}_p^n, \mathsf{F}(E^*)}(G).$$

We sketch the proof, emphasizing how it parallels the proof of the $H^*(G)$ calculation of Section 4. Firstly, χ_G is again constructed using

$$\mathrm{Hom}(\mathbf{Z}_p^n, G)/G \longrightarrow \mathrm{Hom}_{E^*}(E^*(BG), E^*(B\mathbf{Z}_p^n)),$$

where by $E^*(B\mathbf{Z}_p^n)$ we mean $\mathrm{colim}_N\, E^*(B(\mathbf{Z}/p^N)^n)$. Adjointing yields a natural map

$$E^*(BG) \longrightarrow C_{\mathbf{Z}_p^n, E^*(B\mathbf{Z}_p^n)}(G).$$

Formal group law theory yields a map of E^*-algebras

$$\theta \colon E^*(B\mathbf{Z}_p^n) \longrightarrow \mathsf{F}(E^*)$$

which, when composed with the above, defines

$$\chi_G \colon E^*(BG) \longrightarrow C_{\mathbf{Z}_p^n, \mathsf{F}(E^*)}(G).$$

To check that χ_G commutes with induction maps, we use an analogue of Lemma 4.2.

5.3 Lemma [HKR]. *If A' is a proper subgroup of an abelian p-group A, the composite*

$$E^*(BA') \xrightarrow{\text{ind}} E^*(BA) \xrightarrow{\alpha^*} E^*(B\mathbf{Z}_p^n) \xrightarrow{\theta} \mathsf{F}(E^*)$$

is zero, for all surjective homomorphisms $\alpha: \mathbf{Z}_p^n \to A$.

The proof that χ_G is an isomorphism (after extending scalars) then goes as before. The map χ_A is explicitly seen to be an isomorphism if A is abelian — this is the fact corresponding to the Adams-Gunawardena-Miller Theorem. Our generalized Artin's theorem then takes the place of Quillen's F-isomorphism theorem, and thus χ_G can be identified with $\lim\limits_{\mathcal{A}(G)} \chi_A$.

We now use Theorem 5.2 to obtain a slight strengthening of another result in [HKR]:

$$\dim_{K(n)^*} K(n)^{\text{even}}(BG) - \dim_{K(n)^*} K(n)^{\text{odd}}(BG) = |\text{Hom}(\mathbf{Z}_p^n, G)/G|.$$

The strengthening goes as follows. Note that both $G \mapsto K(n)^{\text{even}}(BG)$ and $G \mapsto K(n)^{\text{odd}}(BG)$ are Mackey functors to the category of $K(n)^*$-modules. So one can view $K(n)^{\text{even}}(BG) - K(n)^{\text{odd}}(BG)$ as a "virtual Mackey functor". Call this $\chi_n(G)$

5.3 Theorem. *$[\chi_n(G)] = [C_{\mathbf{Z}_p^n, K(n)^*}(G)]$ in the Grothendieck ring of virtual Mackey functors.*

By this statement, we just mean that for all G, $[\chi_n(G)] = [C_{\mathbf{Z}_p^n, K(n)^*}(G)]$ as virtual representations of the algebra $A(G,G) \otimes \mathbf{Z}/p$ (and so also as virtual $\mathsf{F}_p[\text{Out}(G)]$-modules).

To prove Theorem 5.3, we apply the previous theorem to a specific theory. Using the Bass-Sullivan construction, one can construct a theory E^* with coefficients $\mathbf{Z}_p[v_n, v_n^{-1}]$. The theory E^* can be given a product [Mo][(1)]. Furthermore, there will be a cofibration sequence

$$E \xrightarrow{p} E \longrightarrow K(n),$$

and thus an associated Bockstein spectral sequence.

By Theorems 5.1 and 5.2, $E^*(BG)/(\text{torsion})$ embeds in $C_{\mathbf{Z}_p^n, \mathsf{F}(E^*)}(G)$ as a maximal E^*-lattice. Another such lattice is given by $C_{\mathbf{Z}_p^n, E^*}(G)$. By standard arguments in the theory of modular representations (as in [S, page 125]), one can conclude that

$$[E^*(BG)/(\text{torsion}) \otimes_{E^*} K(n)^*] = [C_{\mathbf{Z}_p^n, E^*}(G) \otimes_{E^*} K(n)^*]$$

as virtual Mackey functors. Note that

$$C_{\mathbf{Z}_p^n, E^*}(G) \otimes_{E^*} K(n)^* = C_{\mathbf{Z}_p^n, K(n)^*}(G).$$

(1) I would like to thank J. Morava for assuring me that this is still true when $p = 2$.

Now we use the Bockstein spectral sequence, with $E^1 = K(n)^*(BG)$, and $E^\infty = E^*(BG)/(\text{torsion}) \otimes_{E^*} K(n)^*$, the latter all in even degrees. Homology preserves Euler characteristics, so $\chi(E^r) = \chi(E^{r+1})$, for all r, as virtual Mackey functors. We conclude that

$$[\chi_n(G)] = [\chi(E^1)] = [\chi(E^\infty)] = [C_{\mathbf{Z}_p^n, K(n)^*}(G)],$$

as needed.

Our analogue of Theorem 4.3 is the following.

5.4 Theorem. *Let G be a p-group of rank d, and let e be an idempotent element in either $\mathbf{F}_p[\mathrm{Out}(G)]$ or $A(G,G) \otimes \mathbf{F}_p$. Then $e\,C_{\mathbf{Z}_p^n,\mathbf{F}_p}(G) = 0$ implies that $e\,K(n)^{\mathrm{odd}}(BG) \neq 0$ for some n. In particular, if there exists a simple $\mathbf{F}_p[\mathrm{Out}(G)]$ or $A(G,G) \otimes \mathbf{F}_p$ module not occurring as a composition factor in $C_{\mathbf{Z}_p^n,\mathbf{F}_p}(G)$, then $K(n)^*(BG)$ is not concentrated in even degrees*(1).

Proof. By the last theorem, $e\,C_{\mathbf{Z}_p^n,\mathbf{F}_p}(G) = 0$ implies that for all n, $\dim eK(n)^{\mathrm{even}}(BG) - \dim eK(n)^{\mathrm{odd}}(BG) = 0$. If $e\,K(n)^{\mathrm{odd}}(BG) = 0$, we would have $0 = e\,K(n)^*(BG) = K(n)^*(eBG)$, for all n. Here eBG denotes the stable retract of $\Sigma^\infty BG_+$ split off by e. We claim that this is impossible, i.e., *no* retract of $\Sigma^\infty BG_+$ can be $K(n)^*$-acyclic for all n. To see this we use the concept of "harmonic spectra" from [R]. By definition, a spectrum Y is harmonic if $[X,Y] = 0$ for all spectra X such that $K(n)_*(X) = 0$ for all n. Any retract of a harmonic spectrum will be harmonic, and if Y is harmonic and not contractible, it follows that $K(n)^*(Y) \neq 0$ for some n. Our proof is completed by the next lemma, due to Mike Hopkins.

5.5 Lemma. *$\Sigma^\infty BG_+$ is harmonic, for all finite groups G.*

Proof. By transfer arguments, we can assume that G is a p-group. In [R], it is shown that any finite complex is harmonic. It is formal that if Y is harmonic, so is any function spectrum $F(Z,Y)$. Thus Spanier-Whitehead duals, $F(Z,S^0)$'s are harmonic. By the Segal conjecture [C2], $\Sigma^\infty BG_+$ is a retract of its own dual.

Now, of course, we wish to find a group G satisfying the criterion of Theorem 5.4, analogous to Example 4.4. We have been unable to find one. The next proposition suggests where one should look for a G.

5.6 Proposition. *Suppose G is a p-group having a simple $\mathbf{F}_p[\mathrm{Aut}(G)]$ module not appearing in $\mathrm{Map}(\mathrm{Hom}(\mathbf{Z}_p^n, G), \mathbf{F}_p)$ for all n. Then G contains a characteristic "class two" subgroup with the same property.*

By class two, we mean a group H such that $H/Z(H)$ is abelian.

(1) (added in proof) John Thompson has shown me that *no* such $\mathrm{Out}(G)$-module can exist.

Proof. Our hypothesis on G is equivalent to the existence of a nonzero idempotent $e \in \mathsf{F}_p[\mathrm{Aut}(G)]$ such that $e\,\mathrm{Map}(\mathrm{Hom}(\mathsf{Z}_p^n, G), \mathsf{F}_p) = 0$ for all n. In the famous Feit-Thompson odd order paper, J. Thompson proved that every finite p-group G contains a characteristic class two subgroup H such that all p'-elements of $\mathrm{Aut}(G)$ act nontrivially on H [G, page 185]. It follows that there are no idempotents in the kernel of $\mathsf{F}_p[\mathrm{Aut}(G)] \to \mathsf{F}_p[\mathrm{Aut}(H)]$, in particular, e projects to a nonzero element of $\mathsf{F}_p[\mathrm{Aut}(H)]$. Since $\mathrm{Map}(\mathrm{Hom}(\mathsf{Z}_p^n, G), \mathsf{F}_p) \to \mathrm{Map}(\mathrm{Hom}(\mathsf{Z}_p^n, H), \mathsf{F}_p)$ is surjective, $e\,\mathrm{Map}(\mathrm{Hom}(\mathsf{Z}_p^n, H), \mathsf{F}_p) = 0$ for all n, also.

5.7 Remark. In fact, H can be chosen so that $H/Z(H)$ is elementary abelian. It is unclear whether $Z(H)$ can also be taken to be elementary.

6 Equivariant cohomology theories

A theme familiar to topologists is introduced in this section. Our character rings are trying to be the coefficients of equivariant cohomology theories.

We need the following construction.

6.1 Definition. Let Γ and G be topological groups, with G compact Lie. Define functors

$$\mathcal{F}_\Gamma(G; \bullet) \colon G\text{-CW-complexes} \to G\text{-CW-complexes}$$

by $\mathcal{F}_\Gamma(G; X) = \{(\alpha, x) \in \mathrm{Hom}(\Gamma, G) \times X \mid \alpha(\gamma)x = x \text{ for all } \gamma \in \Gamma\}$. This is a sub G-space of $\mathrm{Hom}(\Gamma, G) \times X$, where G acts diagonally on the product. Note that if $\mathrm{Hom}(\Gamma, G)$ is finite, then

$$\mathcal{F}_\Gamma(G; X) = \coprod_{\alpha \in \mathrm{Hom}(\Gamma, G)} X^{\mathrm{Im}(\alpha)}.$$

We list some basic properties of these functors.

6.2 Proposition.

(i) $\mathcal{F}_\Gamma(G; *) = \mathrm{Hom}(\Gamma, G)$.

(ii) $\mathcal{F}_\Gamma(G; \bullet)$ *preserves G-pushouts.*

(iii) *If $H < G$, and X is an H-space, there is a natural G-homeomorphism*

$$G \times_H \mathcal{F}_\Gamma(H; X) \simeq \mathcal{F}_\Gamma(G; G \times_H X).$$

(iv) *There is a natural G-homeomorphism*

$$\operatorname*{colim}_{\mathcal{J}\Gamma(G)} \mathcal{F}_\Gamma(H; X) \simeq \mathcal{F}_\Gamma(G; X).$$

(v) $\mathcal{F}_\Gamma(G; \mathcal{F}_\Lambda(G; X)) = \mathcal{F}_{\Gamma \times \Lambda}(G; X)$.

All of these can be verified in a straightforward manner. Checking (v) is a recommended exercise.

Properties (i)–(iii) imply the following, letting G run through finite groups.

6.3 Corollary. *Let E^* be a (non-equivariant) cohomology theory, and F a characteristic zero field. The assignment*

$$X \mapsto E^*(F_\Gamma(G;X))^G \otimes \mathsf{F}.$$

is an equivariant cohomology theory defined for finite G-CW-complexes X.

Since $\mathrm{Hom}(\Gamma, G)$ is discrete, the coefficients $E^*(\mathrm{Hom}(\Gamma,G))^G \otimes \mathsf{F}$ can be identified with $C_{\Gamma,E^*\otimes\mathsf{F}}(G)$. Furthermore, it is easy to extend the formula for induction.

As an example of how one uses such a theory, we prove

6.4 Theorem. *If G is a finite group, and X is a finite G-CW-complex, there is a natural isomorphism*

$$\chi_G \colon K_G(X) \otimes \mathsf{C} \longrightarrow K(F_{\mathsf{Z}}(G;X))^G \otimes \mathsf{C}.$$

Proof. We need to define χ_G. A homomorphism $\alpha\colon \mathsf{Z} \to G$ factors as $\mathsf{Z} \twoheadrightarrow \mathsf{Z}/n \rightarrowtail G$ for some n. The inclusion of Z/n-spaces $X^{\mathsf{Z}/n} \subset X$ induces $K_G(X) \to K_{\mathsf{Z}/n}(X^{\mathsf{Z}/n})$. Now recall that $K_H(Y) = K(Y) \otimes R(H)$ if Y is a trivial H-space [Seg] and that $R(\mathsf{Z}/n) = \mathsf{Z}[x]/(x^n - 1)$. The component of χ_G landing in $K(X^{\mathrm{Im}(\alpha)}) \otimes \mathsf{C}$ is the composite

$$\begin{aligned} K_G(X) \longrightarrow K_{\mathsf{Z}/n}(X^{\mathrm{Im}(\alpha)}) = K(X^{\mathrm{Im}(\alpha)}) \otimes \mathsf{Z}[x]/(x^n - 1) \\ \longrightarrow K(X^{\mathrm{Im}(\alpha)}) \otimes \mathsf{C}, \end{aligned}$$

where the last map is defined by sending x to $e^{2\pi i/n}$.

The proof that χ_G is an isomorphism is immediate. When $X = *$, χ_G reduces to the usual character map

$$R(G) \otimes \mathsf{C} \longrightarrow C_{\mathsf{Z},\mathsf{C}}(G).$$

This is an isomorphism, and one inducts on the cells of X.

6.5 Examples. (1) If $G = \mathsf{Z}/3$, the theorem says that

$$K_{\mathsf{Z}/3}(X) \otimes \mathsf{C} = [K(X) \oplus K(X^{\mathsf{Z}/3}) \oplus K(X^{\mathsf{Z}/3})] \otimes \mathsf{C}.$$

(2) Note that by 6.2(v), $F_{\mathsf{Z},\mathsf{C}}(\mathrm{Hom}(\mathsf{Z}^{n-1}, G)) = \mathrm{Hom}(\mathsf{Z}^n, G)$. Thus the ring $K_G(\mathrm{Hom}(\mathsf{Z}^{n-1}, G))$ is detected by $C_{\mathsf{Z}^n,\mathsf{C}}(G)$ (or, more generally, $C_{\mathsf{Z}^n,\mathsf{F}}(G)$, where F is any char 0 field that contains $|G|$th roots of 1). If G is a p-group, the Atiyah-Segal theorem [AS] implies that $K_G(\mathrm{Hom}(\mathsf{Z}^{n-1}, G))_p = K(EG \times_G \mathrm{Hom}(\mathsf{Z}^{n-1}, G))$.

We leave it to the reader to check that the fact that $EG \times_G G_{\mathrm{conj}} \simeq BG^{S^1}$ (see e.g., [DZ]) implies that $EG \times_G \mathrm{Hom}(\mathbf{Z}^{n-1}, G) \simeq BG^{T^{n-1}}$, when G is finite. Thus $K^*(BG^{T^{n-1}})$ is detected by $C_{\mathbf{Z}^n, \overline{\mathbf{Q}}_p}(G)$, the same ring that detects $E^*(BG)$ when E^* is v_n-periodic. However, in the simple example $n = 2$ and $G = \mathbf{Z}/p$, the ring structures of $K((B\mathbf{Z}/p)^{S^1})$ and $E^*(B\mathbf{Z}/p)$ are different: the former looks roughly like an associated graded of the latter.

7 Simplicial character rings

It has occurred to a number of people who have heard about our character rings for v_n-periodic theories that they should be assembled into some sort of co-simplicial object. Here I wish to advertise a question motivated by this idea.

Recall that BG is the geometric realization of the simplicial set $\underline{n} \mapsto G^n = \mathrm{Hom}(F_n, G)$. This suggests the following constructions.

7.1 Definitions. *Let Γ be a co-simplicial group.*

(i) Let $B_\Gamma G$ be the realization of $\underline{n} \mapsto Hom(\Gamma_n, G)$.

(ii) Let $\overline{B}_\Gamma G$ be the realization of $\underline{n} \mapsto Hom(\Gamma_n, G)/G$.

Note that the chain complex computing $H^*(\overline{B}_\Gamma G; \mathsf{F})$ has n-chains isomorphic to $C_{\Gamma_n, \mathsf{F}}(G)$. This looks promising. However, this construction is too naive; induction does not generally commute with our differentials. Still, it is tempting to think that $\overline{B}_{(\mathbf{Z}/p)^\bullet} G$ has something to do with $H^*(G; \mathbf{Z}/p)$ and that $\overline{B}_{\mathbf{Z}^\bullet} G$ has something to do with v_n-periodicity.

7.2 Problem. Find a functor

$$\mathcal{B} : \text{finite groups} \longrightarrow \text{simplicial sets}$$

such that

(i) there is a natural homotopy equivalence $\Sigma^\infty |\mathcal{B}G| \simeq \Sigma^\infty BG$, and

(ii) the chain complex $C^*(\mathcal{B}G; \mathsf{F})$ is a complex of Mackey functors inducing the transfer in cohomology.

The point here is that when one defines the usual stable homotopy transfer for $H < G$, $\Sigma^\infty BG_+ \to \Sigma^\infty BH_+$, one "trades in" BH for $EG \times_G (G/H)$, a homotopic, but different, space.

References

[AGM] J.F. Adams, J.H.C. Gunawardena, and H.R. Miller: The Segal conjecture for elementary abelian p-groups. *Topology* 24 (1985), 435–460.

[A] M.F. Atiyah: Characters and cohomology of finite groups. *Inst. Hautes Études Sci. Publ. Math.* 9 (1961), 23–64.

[AS] M.F. Atiyah and G.B. Segal: Equivariant K-theory and completion. *J. Differential Geom.* 3 (1969), 1–18.

[C1] G. Carlsson: G.B. Segal's Burnside ring conjecture for $(\mathbb{Z}/2)^k$. *Topology* 22 (1983) 83–103.

[C2] G. Carlsson: Equivariant stable homotopy and Segal's Burnside ring conjecture. *Ann. of Math.* 120 (1984), 189–224.

[DZ] W. Dwyer and A. Zabrodsky: Maps between classifying spaces. *Lecture Notes in Math. 1298, pp. 106–119, Springer*, 1987.

[G] D. Gorenstein: Finite Groups. Harper and Row, New York, 1968.

[HKR] M.J. Hopkins, N.J. Kuhn, and D.C. Ravenel: Generalized group characters and complex oriented cohomology theories. In preparation.

[HLS] H.-W. Henn, J. Lannes, and L. Schwartz: Analytic functors, unstable algebras and cohomology of classifying spaces. Preprint, 1988.

[K1] N.J. Kuhn: The mod p K-theory of classifying spaces of finite groups. *J. Pure Appl. Algebra* 44 (1987), 269–271.

[K2] N.J. Kuhn: The Morava K-theories of some classifying spaces. *Trans. Amer. Math. Soc.* 304 (1987), 193–205.

[LS] J. Lannes and L. Schwartz: Sur la structure des A-modules instable injectifs. *Topology* 28 (1989), 153–169.

[Ma] J.P. May: Stable maps between classifying spaces. *Contemp. Math.* 37 (1985), 121–129.

[Mi] H.R. Miller: The Sullivan conjecture on maps from classifying spaces. *Ann. of Math.* 120 (1984), 39–97.

[Mo] J. Morava: A product for the odd-primary bordism of manifolds with singularities. *Topology* 18 (1979), 177–186.

[MP] S.A. Mitchell and S.B. Priddy: Symmetric product spaces and splittings of classifying spaces. *Amer. J. Math.* 106 (1984), 219–232.

[N] S. Norton: Generalized moonshine. *Appendix to G. Mason: Finite groups and modular functions, Proc. Sympos. Pure Math.* 47 (1987), 181–210.

[Q1] D. Quillen: The spectrum of an equivariant cohomology ring. *Ann. of Math.* 94 (1971), 549–572.

[Q2] D. Quillen: The Adams conjecture. *Topology* 10 (1971), 67–80.

[R] D.C. Ravenel: Localization with respect to certain periodic homology theories. *Amer. J. Math.* 106 (1984), 351–414.

[S] J.-P. Serre: Linear Representations of Finite Groups. Springer, New York, 1977.

[Se] G. Segal: Equivariant K-theory. *Inst. Hautes Études Sci. Publ. Math.* 34 (1968), 129–151.

The rank filtration of Quillen's Q-construction

Stephen A. Mitchell*

Introduction

The Quillen K-groups of a ring R are defined by $K_iR = \pi_{i+1}BQ\mathcal{P}R$ [Q2]. Here $Q\mathcal{P}R$ is a certain category whose objects are finitely generated projective R-modules, and $BQ\mathcal{P}R$ is its classifying space. If R is commutative then $Q\mathcal{P}R$, and hence $BQ\mathcal{P}R$, can be filtered by the rank of a module. This filtration was exploited by Quillen [Q3] to show that the K-groups of a ring of integers R in an algebraic number field are finitely generated. The key step in his argument is the calculation of the homology of the quotients of the rank filtration. These homology groups are computed in terms of the homology of the Steinberg module St; for example, if R is a principal ideal domain then

$$H_*(k\text{th quotient}) = H_{*-k}(GL_kR, St_k).$$

Since St_k is by definition $\widetilde{H}_{k-2}(B(R^k))$, where $B(R^k)$ — the "building" or flag complex — is equivalent to a wedge of $(k-2)$-spheres, this suggests that the kth quotient is actually equivalent to the reduced Borel construction $EGL_kR^+ \wedge_{GL_kR} \Sigma^2 B(R^k)$. In Section 1 of this paper we prove this (presumably well-known) fact in a very general form. (In fact we don't consider the Q-construction itself, but an equivalent construction due to Waldhausen [W1].)

The motivation for this work was a question raised by Gunnar Carlsson: Does the rank filtration stably split (for at least some rings R, after suitable localization)? When R is a ring of integers in a number field, such a stable splitting would be useful in Carlsson's program [Ca] for attacking the Lichtenbaum-Quillen conjectures: it would facilitate passing from results about GL_kR to results about GLR. The question seems reasonable since both D. Kahn's splitting of QS^1 [K] and H. Miller's splitting of U [Mil] can be viewed as examples of splitting the rank filtration (in the context of Waldhausen's version of the Q-construction). This observation is due to

*Partially funded by the National Science Foundation

Carlsson, and is explained in Section 2. In fact one can even give parallel proofs of these splittings (see Section 3). Unfortunately, however, it seems unlikely that there is any general result of this nature. For it is easy to show (Section 2) that U/O does not split (this would be the real analogue of Miller's theorem; it may well split at odd primes). More discouraging is the fact that when R is a finite field F_q, there is no splitting at all, even after localization (Section 4). However the results of Section 4 may be of some independent interest. We calculate (1) $H_*BQ\mathcal{P}\mathsf{F}_q$ as a Hopf algebra over the Steenrod algebra (at a noncharacteristic prime ℓ) and (2) $H_*(GL_k\mathsf{F}_q, St_k)$. It follows that the rank filtration on homology does at least split additively.

I would like to thank Gunnar Carlsson for explaining his ideas on this subject to me.

1 The rank filtration of the Q-construction, and its quotients

Let R be a ring. Quillen initially defined the higher K-groups of R in terms of his "plus construction": $K_iR = \pi_iBGLR^+$ for $i > 0$. On the other hand one really should think of the K-groups as invariants of the *category* $\mathcal{P}R$ of finitely generated projective modules over R. To define higher K-groups for other categories of interest — e.g., all finitely generated modules over R, or vector bundles or coherent sheaves on an algebraic variety — a different approach is needed. This was also provided by Quillen, in his famous paper [Q2]. Starting with an "exact category" $\mathcal{C}$ (such as $\mathcal{P}R$), Quillen defines a new category $Q\mathcal{C}$. He then defines the K-groups of $\mathcal{C}$ by $K_i\mathcal{C} = \pi_{i+1}BQ\mathcal{C}$, where $BQ\mathcal{C}$ is the classifying space of $Q\mathcal{C}$. This is the Q-construction. The agreement of the two definition in the case $\mathcal{C} = \mathcal{P}R$ is Quillen's "Q equals plus" theorem: $\Omega BQ\mathcal{P}R \cong BGLR^+ \times K_0R$ (see Grayson [G]).

The Q-construction by definition applies only to exact categories. For example, it does not apply to the category of finite sets. On the other hand the Barratt-Priddy-Quillen theorem [P] says that $B\Sigma_\infty^+ \cong QS^0$; here $Q = \Omega^\infty\Sigma^\infty$. Thus the stable homotopy groups of spheres are the K-groups of the category of finite sets, and one would like a more general construction that yields a delooping of QS^0 in this case. One approach (with many other applications) is due to Waldhausen. It applies to any category "with cofibrations and weak equivalences". We will describe here a simplified version in which the weak equivalences are the isomorphisms. A *category with cofibrations* is a category $\mathcal{C}$ with a null object $*$ and a subcategory $co\mathcal{C}$ of "cofibrations" such that (a) isomorphisms are cofibrations (b) $* \to W$ is a cofibration for all W and (c) $co\mathcal{C}$ is closed under cobase change. Here (c) means that for any diagram

$$\begin{array}{ccc} W_0 & \xrightarrow{i} & W_1 \\ \downarrow & & \\ W_2 & & \end{array}$$

with i a cofibration, the pushout W exists and the resulting map $W_2 \to W$ is again a cofibration. In particular, taking $W_2 = *$, every cofibration $W_0 \to W_1$ has a "cofibre" W_1/W_0, unique up to isomorphism.

Examples. (i) Finitely generated projective R-modules/split monomorphisms, (ii) finitely generated free R-modules/monomorphisms with free quotient, (iii) more generally any exact category/admissible monomorphisms, (iv) pointed finite sets/pointed injections.

Define a new category $\mathcal{C}_n$ whose objects are sequences of cofibrations

$$* \longrightarrow W_1 \longrightarrow \ \dots \ \longrightarrow W_n$$

and morphisms the commutative diagrams

$$\begin{array}{ccccc} * \longrightarrow & W_1 & \longrightarrow & \dots & \longrightarrow W_n \\ & \cong\downarrow & & & \downarrow\cong \\ * \longrightarrow & W_1' & \longrightarrow & \dots & \longrightarrow W_n' \end{array}$$

in which the vertical maps are isomorphisms.

Then $n \mapsto \mathcal{C}_n$ defines (almost) a simplicial category: degeneracies are obtained by inserting identity maps in the obvious way and face maps d_i, $1 \le i \le n$, by deleting W_i. On the other hand d_0 is special. On objects it takes $* \to W_1 \to \dots \to W_n$ to $* \to W_2/W_1 \to \dots \to W_n/W_1$. (There is a minor technical problem here involving the choice of cofibres W_i/W_1, see [W]. In order to get a simplicial category, $\mathcal{C}_n$ must be replaced by an equivalent category which has all possible choices built in. This technicality will be suppressed in the sequel.) Taking classifying spaces yields a simplicial space $W_\bullet\mathcal{C}$, and Waldhausen's generalized Q-construction is the geometric realization $|W_\bullet\mathcal{C}|$.

Examples. (1) If $\mathcal{C}$ is an exact category, then (Quillen, Waldhausen) $|W_\bullet\mathcal{C}| \cong BQ\mathcal{C}$.

(2) $\mathcal{C} = \mathrm{Set}_* =$ finite pointed sets. Then $|W_\bullet\mathcal{C}|$ is clearly equivalent to BM, the classifying space of the monoid $M = \coprod_{k\geq 0} B\Sigma_k$. It is well known that $BM \cong QS^1$ (here $Q = \Omega^\infty\Sigma^\infty$). In fact it is an amusing exercise to show this explicitly, using a configuration space model of QS^1.

(3) $\mathcal{C} = \mathcal{F}R$, where R is a ring with the invariant dimension property — i.e., the rank of a finitely generated free module is well-defined. It is then clear that $(\mathcal{F}R)_n$ is equivalent to the subcategory of sequences of inclusions

$$0 \subseteq R^{i_1} \subseteq R^{i_2} \dots \subseteq R^{i_n},$$

where $0 \le i_1 \le \dots \le i_n$ and the inclusions $R^i \subseteq R^j$ are defined as usual in terms of the standard ordered basis; the morphisms are the automorphisms. Let $I = (i_1, \dots, i_n)$

and let P_I denote the automorphism group of the corresponding sequence. Then P_I is a standard parabolic subgroup of $GL_{i_n}R$. Clearly the simplicial space $W_\bullet \mathcal{F}R$ is equivalent to the evident simplicial space $W_\bullet R$, where $W_n R = \coprod_{\ell(I)=n} BP_I$. For example, if $0 < j < n$ then $d_j : BP_I \to BP_{I'}$ is induced by the natural inclusion $P_I \subseteq P_{I'}$, where I' is obtained from I by deleting the jth entry. On the other hand d_0 (respectively d_n) involves projection on $P_{I'}$ where $I' = (i_2 - i_1, \ldots, i_n - i_1)$ (respectively $(i_1, \ldots, i_{n-1})$). Note that since $|W_\bullet \mathcal{F}R|$ is an H-space with $\pi_1 \cong \mathbf{Z}$, it splits as $S^1 \times$ (universal cover).

A category with cofibrations $\mathcal{C}$ is *ranked* if it is equipped with a rank function $r : Ob\mathcal{C} \to \mathbf{Z}^+$ such that, if $i : W_1 \to W_2$ is a cofibration, then (a) $r(W_2) = r(W_1) + r(W_2/W_1)$ (i.e., r is additive), and (b) i is an isomorphism if $r(W_1) = r(W_2)$. The obvious examples are the usual rank on $\mathcal{P}R$, if R is commutative; the usual rank on $\mathcal{F}R$, if R has the invariant dimension property; and $r =$ (cardinality -1) for Set$_*$. Or one could take $\mathcal{C} =$ finitely generated modules over an Artinian ring, $co\mathcal{C} =$ monomorphisms, and $r(M) =$ number of composition factors of M. It is clear that rank induces a filtration of the simplicial category $\mathcal{C}$. Let $\mathcal{C}_n^k =$ full subcategory of $\mathcal{C}_n$ consisting of sequences with $r(W_n) \leq k$. Then $\mathcal{C}_\bullet^k$ is a sub-simplicial category, and we obtain a filtration $W_\bullet^k \mathcal{C}$ of the simplicial space $W_\bullet \mathcal{C}$.

In order to describe the quotients of the filtration we need to introduce the "building" $B(V)$ associated to a fixed object V. Regard V as an object of $co\mathcal{C}$, and let $P(V)$ denote the poset of sub-objects of V. In view of our axioms for the rank function, it's clear that two cofibrations $W_0 \to V$, $W_1 \to V$ represent the same sub-object if and only if there is an isomorphism $W_0 \to W_1$ over V. If we assume cofibrations are monic, this isomorphism is unique. Now let $P_0(V)$ denote the sub-poset (possibly empty) of sub-objects W with $0 < r(W) < r(V)$. Then $B(V)$ is defined to be the nerve of $P_0(V)$; it is a simplicial Aut V-set.

Examples. (1) $\mathcal{C} =$ Set$_*$, $V = \{1, 2, \ldots, n\}$. Then $B(V) \cong S^{n-2}$.

(2) $\mathcal{C} = \mathcal{P}R$, R a division ring. Then (Solomon-Tits) $B(R^n)$ is equivalent to a wedge of $(n-2)$-spheres (see [Q3] for a proof). The Steinberg module St_n is $\widetilde{H}_{n-2}(B(R^n); \mathbf{Z})$, regarded as a $GL_n R$-module ($n \geq 2$). For $n = 1$ we set $St_1 = \mathbf{Z}$, with trivial action.

(3) $\mathcal{C} = \mathcal{P}A$, A a Dedekind domain with quotient field F. If P is a projective A-module of rank n, then $P \to P \otimes_A F \equiv V$ induces an isomorphism of posets $B(P) \to B(V)$ (this follows easily from the fact that a finitely generated module over a Dedekind domain is projective if and only if it is torsion-free). In particular $B(P)$ is equivalent to a wedge of S^{n-2}'s.

It is convenient to assume that $\mathcal{C}$ has two further properties: (a) cofibrations are monic, and (b) for each object V, the functor $\Pi : co\mathcal{C}/V \to P(V)$ taking an object to its equivalence class has a right inverse $S : P(V) \to co\mathcal{C}/V$, with $S(V) = 1_V$. We then say that $\mathcal{C}$ "has inclusions". Then if $W_0 \xrightarrow{i} V$ is an inclusion (i.e., $i \in \operatorname{Im} S$), and $g \in \operatorname{Aut} V$, there is a unique inclusion $W_1 \xrightarrow{j} V$ and a unique isomorphism $\varphi : W_0 \to W_1$ such that

$$\begin{array}{ccc} W_0 & \xrightarrow{i} & V \\ \varphi\downarrow & & \downarrow g \\ W_1 & \xrightarrow{j} & V \end{array}$$

commutes. Thus $B(V)$ is the nerve of the poset of proper inclusions. The $\operatorname{Aut} V$ action is then described as follows. If $W_1 \to W_2 \to \ldots \to W_p = V$ is a chain of inclusions, and $g \in \operatorname{Aut} V$, we get:

1.1 *There is a unique chain of inclusions $W_1' \to W_2' \to \ldots \to W_p' = V$ and there are unique isomorphisms $W_i \to W_i'$ such that*

$$\begin{array}{ccccccccc} W_1 & \longrightarrow & W_2 & \longrightarrow & \cdots & \longrightarrow & W_p & = & V \\ \downarrow & & \downarrow & & & & \downarrow & & \downarrow g \\ W_1' & \longrightarrow & W_2' & \longrightarrow & \cdots & \longrightarrow & W_p' & = & V \end{array}$$

commutes.

In fact we will need the double suspension of $B(V)$, denoted $\widetilde{B}(V)$. Since $B(V)$ typically has no $\operatorname{Aut} V$-fixed points, we must take the unreduced suspension first, followed by reduced suspension. To be quite explicit, recall the following construction of the unreduced suspension of a poset. Suppose P is a poset with minimal element 0 and maximal element V, and let $P_0 = P - \{0, V\}$. Then we can form a new simplicial set $X_\bullet$ with X_p =chains $W_0 \leq W_1 \ldots \leq W_p$ with $W_i \in P$, subject to the condition $W_0 = 0 \Longrightarrow W_p \neq V$. Obviously $|X_\bullet|$ is the unreduced suspension of $|P_0|$. Alternatively, we can factor out the cone consisting of all chains with $W_p \neq V$, obtaining an equivalent simplicial set $\widetilde{\Sigma} P_0$. This will be our model for the unreduced suspension. If P is a G-poset, then the two vertices $0, V$ of $\widetilde{\Sigma} P_0$ are fixed by G. Taking 0 as the basepoint, we can now form a reduced suspension $\Sigma\widetilde{\Sigma} P_0$. This has p-simplices the chains $W_1 \leq \ldots \leq W_p$ with $W_p = V$, together with the unique degeneracy of the basepoint $*$ in degree zero. Note that d_0 and d_p map any nondegenerate simplex to the basepoint. If $n \geq 2$ we then define $\widetilde{B}_n(V) = \Sigma\widetilde{\Sigma} B_n(V)$; this is a simplicial pointed $\operatorname{Aut} V$-set. We set $\widetilde{B}(*) =$ point. If rank $V > 0$ and $B(V) = \emptyset$, we set $\widetilde{B}(V) =$ usual simplicial model of S^1, with just two nondegenerate simplices (and trivial $\operatorname{Aut} V$ action).

1.2 **Theorem.** *Let $\mathcal{C}$ be a ranked category with cofibrations and inclusions. Then, for $k \geq 1$,*

$$|W_\bullet^k \mathcal{C}| / |W_\bullet^{k-1} \mathcal{C}| \cong \bigvee_{r(V)=k} E\operatorname{Aut} V^+ \wedge_{\operatorname{Aut} V} |\widetilde{B}(V)|$$

where the wedge is over isomorphism classes of objects of rank k.

Proof. First recall that any Borel construction is the realization of a simplicial category. If G is a group and Y is a G-set, let $T(Y)$ denote the translation category of Y. Thus $T(Y)$ has objects the points of Y, while a morphism $y_1 \to y_2$ is an element $g \in G$ with $gy_1 = y_2$. The nerve $NT(Y)$ is just the usual simplicial model (the bar construction) for $EG \times_G Y$. Now suppose $X_\bullet$ is a simplicial G-set. Since $Y \mapsto T(Y)$ is a functor (G-sets) $\to$ (categories), $T(X_\bullet)$ is a simplicial category. Furthermore its realization is $EG \times_G |X|$. Form the evident bisimplicial set X_{pq} with $X_{pq} = G^p \times X_q$. (That is, $X_\bullet$ is the simplicial simplicial set $q \mapsto NT(X_q)$.) Realizing in the p-direction first yields $|T(X_\bullet)|$, while realizing in the q-direction first yields $EG \times_G |X|$; this proves the claim. Thus $E\mathrm{Aut}\, V \times_{\mathrm{Aut}\, V} |\widetilde{B}(V)|$ is the realization of the simplicial category $p \mapsto T(\widetilde{B}(V)_p)$. Explicitly, for $p > 0$ an object of this category is a chain of inclusions

$$* \longrightarrow W_1 \longrightarrow \dots \longrightarrow W_{p-1} \longrightarrow W_p$$

where either $W_p = V$ or $W_p = *$ with morphisms the elements of $\mathrm{Aut}\, V$ as above. When $p = 0$ we get the group (category) $\mathrm{Aut}\, V$.

Now consider $|W_\bullet^k \mathcal{C}|/|W_\bullet^{k-1}\mathcal{C}| = |W_\bullet^k \mathcal{C}/W_\bullet^{k-1}\mathcal{C}|$. Clearly this is equal to a wedge $\bigvee_{r(V)=k} |\mathcal{C}_\bullet(V)|$ where V ranges over a set of representatives of isomorphism classes of objects of rank k, and $\mathcal{C}_\bullet(V)$ is the following simplicial category: $\mathcal{C}_p(V)$ has objects the chains of cofibrations

$$W_0 \longrightarrow W_1 \longrightarrow \dots \longrightarrow W_p$$

with $W_0 = *$ and either $W_p \cong V$ or $W_p = *$. The morphisms are the isomorphisms as before. Note $\mathcal{C}_p(V)$ has a basepoint (all $W_i = *$), which will itself be denoted $*$. The simplicial operators are as in $\mathcal{C}_\bullet$, except that now $d_0 = *$ if $r(W_1) \neq 0$, and $d_p = *$ if $r(W_{p-1}) < k$. Define a functor $F_p : T(\widetilde{B}(V))_p \to \mathcal{C}_p(V)$ via the obvious inclusion if $p > 0$. (On morphisms, F_p is defined using 1.1.) Since $\mathcal{C}_0(V)$ is the trivial category with one object $*$, F_0 is of course the constant functor. Clearly $F_\bullet$ is a map of simplicial categories, and so induces a map

$$|F| : E\mathrm{Aut}\, V \times_{\mathrm{Aut}\, V} |\widetilde{B}(V)| \longrightarrow |\mathcal{C}_\bullet(V)|,$$

which factors through a map $H : E\mathrm{Aut}\, V^+ \wedge_{\mathrm{Aut}\, V} |\widetilde{B}(V)| \to |\mathcal{C}_\bullet(V)|$. Finally, it is easy to check that F_p is an equivalence of categories if $p > 0$. Thus H is the realization of a map of simplicial spaces which is an equivalence on each level, and so is an equivalence. This completes the proof. ∎

1.3 **Corollary** (Quillen). *If F is a division ring or a ring of algebraic integers in a number field, there is a first quadrant spectral sequence*

$$E^1_{p,q} = \bigoplus_{\mathrm{rank}\, V = p} H_q(\mathrm{Aut}\, V, St) \Longrightarrow H_{p+q} BQPR.$$

Proof. This is just the spectral sequence of the filtered space $|W_\bullet\mathcal{P}R|$. Since $\widetilde{B}(V)$ is $(p-1)$-connected, it follows that we get finite convergence, and $H_{p+q}(E\mathrm{Aut}\,V^+ \wedge_{\mathrm{Aut}\,V} \widetilde{B}(V)) \cong H_q(\mathrm{Aut}\,V, St)$.

1.4 Corollary (Quillen). *If F_q is a finite field of characteristic p,*

$$(K_i\mathsf{F}_q)\otimes \mathbb{Z}_{(p)} = 0 \quad \text{for } i > 0.$$

Proof. For $n \geq 2$, the mod p Steinberg representation $St \otimes \mathbb{Z}/p$ of $GL_n\mathsf{F}_q$ is nontrivial, projective and irreducible. Hence $H_i(GL_n\mathsf{F}_q, St\otimes\mathbb{Z}/p) = 0$ for all $i \geq 0$, $n \geq 2$, and 1.4 follows from 1.3.

1.5 Corollary (Quillen). *If A is the ring of integers in an algebraic number field, K_iA is finitely generated for all i.*

Proof. It follows from the work of Borel and Serre [B-S] that for any finitely generated projective P, $H_i(\mathrm{Aut}\,P, St)$ is finitely generated for all i. Since by the Jordan-Zassenhaus Theorem there are only finitely many projectives of fixed rank, 1.5 follows from 1.3. ∎

2 The rank filtration for topological K-theory

Let U denote the infinite unitary group. There is a well-known filtration given by $F_kU = \{A \in U : \operatorname{codim}\{X \in \mathbb{C}^\infty : AX = X\} \leq k\}$. The stratum $F_kU - F_{k-1}U$ is obviously identified with the space of pairs (W, A), where W is a subspace of $\mathbb{C}^\infty$, $\dim W = k$, and A is a unitary automorphism of W with eigenvalues $\neq 1$. This in turn is the total space of the adjoint bundle over $BU(k)$: $F_kU - F_{k-1}U = EU(k)\times_{U(k)} \mathfrak{u}(k)$, where $\mathfrak{u}(k)$ is the adjoint representation of $U(k)$ on its Lie algebra (See for example [Cr].) Thus $F_kU/F_{k-1}U$ is the Thom space $BU(k)^{\mathrm{ad}}$ of the adjoint bundle.

Now Gunnar Carlsson has observed that this filtration can be interpreted as an instance of the rank filtration of Section 1. More precisely one first needs to generalize the results discussed in Section 1 to suitable topological categories. However we are only concerned here with some very simple special cases, so we will proceed on an ad hoc basis. Let $\mathbb{C}^{\mathrm{top}}$ be the topological category of finite-dimensional inner product spaces over $\mathbb{C}$. As cofibrations we take the injective isometries. Then, clearly, $W_\bullet\mathbb{C}^{\mathrm{top}}$ is equivalent to the classifying space BM of the monoid $\coprod_k BU(k)$. One knows that BM is equivalent to U, but it is amusing to exhibit this equivalence in an explicit form: it is nothing more than the eigenspace decomposition of a unitary matrix. If $J = (j_1, \ldots j_n)$ is a sequence of nonnegative integers, let X_J = space of n-tuples $(W_1, \ldots, W_n)$ of pairwise orthogonal subspaces of $\mathbb{C}^\infty$, with $\dim W_s = j_s$. Then $X_J \cong BU(j_1)\times\ldots\times BU(j_n)$. We can define in the evident way a simplicial space $X_\bullet$

with $X_n = \coprod_{\ell(J)=n} X_J$, equivalent to BM. (The face operators d_i are given by direct sum $W_i \oplus W_{i+1}$ if $1 \leq i \leq n-1$, while d_0 (respectively d_n) omits W_1 (respectively W_n).) An explicit homeomorphism $|X_\bullet| \xrightarrow{f} U$ is then given as follows. Write the n-simplex Δ^n as $\{(a_1, \ldots, a_n) \in \mathbf{R}^n \colon 0 \leq a_1 \leq a_2 \ldots a_n \leq 1\}$, with faces given by $a_1 = 0$, $a_i = a_{i+1}, a_n = 1$. Map $X_n \times \Delta^n \xrightarrow{f} U$ by setting $f((W_1, \ldots, W_n), (a_1, \ldots, a_n)) = A$, where A is the identity I on $(W_1 \oplus \ldots \oplus W_n)^\perp$ and $A|_{W_i} = e^{2\pi i a_i} I$. It is easy to check that f induces a well-defined homeomorphism $|X_\bullet| \cong U$. Now the rank filtration corresponds to $|J| \leq k$, where $|J| = j_1 + \ldots + j_n$, so we see that f is filtration preserving.

To complete the discussion we should show that the two descriptions of the quotients of the filtration agree. Since $BU(k)^{\mathrm{ad}} = EU(k)^+ \wedge_{U(k)} S^{\mathrm{ad}}$, where we use the notation S^V for the one-point compactification of a representation V, the point is to show that, as a $U(k)$-space, the suspended topological building $\widetilde{B}^{\mathrm{top}}(\mathbf{C}^n)$ is just S^{ad}. This is in fact a special case of a more general result. If G is a connected semisimple compact Lie group, with complexification $G_{\mathbf{C}}$, there is an associated Tits building $\widetilde{B}(G_{\mathbf{C}})$. This is the realization of a certain simplicial set analogous to the flag complex, and has the homotopy type of an infinite wedge of spheres. On the other hand if one takes into account the topology on $G_{\mathbf{C}}$, the simplicial set becomes a simplicial space, with realization G-homeomorphic to the unit sphere in the adjoint representation of G. (See [Mit] for a detailed discussion.) In particular we can take $G = SU(n)$. From this point of view, the two suspensions involved in the definition of $\widetilde{B}$ arise from (1) replacing the unit sphere by one-point compactification and (2) replacing $SU(n)$ by $U(n)$. In any event, the equivalence $\widetilde{B}^{\mathrm{top}}(\mathbf{C}^n) \cong S^{\mathrm{ad}}$ is easy to see directly, via eigenspace decomposition (see Section 3).

We next consider real topological K-theory. The real analog of U is U/O in this context, since $\Omega(U/O) = BO \times \mathbf{Z}$. Recall that the map $U(n) \to U(n)$ given by $A \to AA^T$ induces a homeomorphism $U(n)/O(n) \cong$ symmetric unitary matrices. Thus $U(n)/O(n)$ is the fixed point set of an involution on $U(n)$, and much of the preceding discussion for $\mathbf{C}^{\mathrm{top}}$ carries over immediately to $\mathbf{R}^{\mathrm{top}}$ by taking fixed points. For instance, U/O inherits a filtration from U, with $F_k(U/O)/F_{k-1}(U/O) = BO(n)^{\mathrm{sym}}$ where sym is the natural representation of $O(n)$ on $n \times n$ real symmetric matrices. Furthermore, $|W_\bullet \mathbf{R}^{\mathrm{top}}| = BM$ where $M = \coprod BO(k)$. As before, we can replace BM by an equivalent simplicial space homeomorphic to U/O via an eigenspace decomposition. (Note that the symmetric unitary matrices are precisely the unitary matrices with real eigenspaces, i.e. if W is an eigenspace then $W = \mathbf{C} \cdot (W \cap \mathbf{R}^n)$.) Again, we see at once that the filtration F_k corresponds to the rank filtration on $W_\bullet \mathbf{R}^{\mathrm{top}}$, and the building $\widetilde{B}^{\mathrm{top}}(\mathbf{R}^n)$ is $O(n)$-equivariantly homeomorphic to S^{sym}. The latter remark is again a special case of a more general fact. If G is a connected semisimple Lie group with involution σ, then σ extends uniquely to an anti-holomorphic involution on $G_{\mathbf{C}}$.

Now $G_{\mathbf{R}} \equiv (G_{\mathbf{C}})^{\sigma}$ is a semisimple real Lie group, and there is an associated topological Tits building $B^{\text{top}}(G_{\mathbf{R}})$. This can be identified as follows. Let $\mathfrak{m}$ denote the (-1)-eigenspace of σ on the Lie algebra $\mathfrak{g}$, regarded as a representation of the fixed group G^{σ}. Then $B^{\text{top}}(G_{\mathbf{R}})$ is G^{σ}-equivariantly homeomorphic to the unit sphere in $\mathfrak{m}$.

The induced filtration on mod 2 homology and cohomology is easily described. Recall $H^*(U/O, \mathsf{F}_2) = \mathsf{F}_2 \langle w_1, w_2, \ldots \rangle$, where the w_i's are the Stiefel-Whitney classes of the map $U/O \xrightarrow{i} BO$ occurring in the fibre sequence $U/O \to BO \to BU$. In fact the Eilenberg-Moore spectral sequence collapses to its edge homomorphism: $\mathrm{Tor}_0^{H^*BU}(\mathsf{F}_2, H^*BO) \cong H^*(U/O)$. Dually, i_* is an isomorphism onto the subalgebra generated by the primitives. Hence $H_*(U/O, \mathsf{F}_2)$ is a polynomial algebra $\mathsf{F}_2[s_1, s_3, \ldots]$ on the odd-dimensional primitives s_i. In fact the inclusion $\Sigma \mathbf{R}P_+^{\infty} \cong F_1(U/O) \subseteq U/O$ induces an isomorphism onto the primitives. Hence the primitives $s_1, s_2, \ldots$ form a simple system of generators, and one can easily check that the rank filtration on H_* is just the length filtration on monomials in the s_i. Dually the rank filtration on H^* is the length filtration on monomials in the w_i. In particular the rank filtration splits *additively* on H^*, i.e. the inclusions $F_{k-1}(U/O) \subseteq F_k(U/O)$ are surjective on cohomology. (Note that the vector space of monomials of length k in $w_1, \ldots, w_k$ has Poincaré series $= t^{\binom{k+1}{2}} \Pi_{i=1}^k (1-t^i)^{-1}$ = Poincaré series of $BO(k)^{\text{sym}}$.) However it does not split over the Steenrod algebra. For in particular this would imply that the module of indecomposables splits off — i.e. there would be a set of exterior generators $x_1, x_2, \ldots$ closed under the A-action. But the Wu formula quickly leads to a contradiction. We have $x_1 = w_1, x_2 = w_2, x_3 = \mathrm{Sq}^1 w_2 = w_3 + w_1 w_2$, $x_5 = \mathrm{Sq}^2 x_3 = w_5 + w_1 w_4 + w_2 w_3$. But any choice of x_4 will satisfy $\mathrm{Sq}^1 x_4 = w_5 + w_1 w_4$.

3 Remarks on stable splittings of the rank filtration

When $\mathcal{C}$ = finite pointed sets or inner product spaces over $\mathbf{C}$, the rank filtration of $|W_\bullet \mathcal{C}|$ stably splits.

3.1 Theorem (D. Kahn). $QS^1 \cong \bigvee_{k=1}^{\infty} E\Sigma_k^+ \wedge_{\Sigma_k} S^k$ *(stably).*

3.2 Theorem (H.Miller [Mil]). $U \cong \bigvee_{k=1}^{\infty} EU(k)^+ \wedge_{U(k)} BU(k)^{\text{ad}}$ *(stably).*

In fact there are very similar proofs of these splittings: for 3.1 we can use Ralph Cohen's elegant argument [Co]. The May-Milgram model for QS^1 has the form $\coprod_k E\Sigma_k \times_{\Sigma_k} (S^1)^k / \sim$ for a certain equivalence relation $\sim$. In particular there is a quotient map $E\Sigma_k \times_{\Sigma_k} (S^1)^k \to F_k QS^1$, where F_k is the obvious filtration, such that the composite

$$E\Sigma_k \times_{\Sigma_k} (S^1)^k \longrightarrow F_k QS^1 \longrightarrow F_k QS^1 / F_{k-1} QS^1 = E\Sigma_k^+ \wedge_{\Sigma_k} S^k$$

is just the obvious map induced by $(S^1)^k \xrightarrow{\pi} S^k$. It would be enough to show that this composite is a stable retraction. But Lewis, May and Steinberger have shown [LMS] that the equivariant half-smash product $EG^+ \wedge_G X$ (G a compact Lie group) can be defined on the category of G-spectra. Hence it is enough to show that π is a stable Σ_k-equivariant retraction. For if $S^k \xrightarrow{s} (S^1)^k$ is a map of Σ_k-spectra with $\pi s \cong 1$, we can simply apply the functor $E\Sigma_k^+ \wedge_{\Sigma_k}$ to get the desired splitting map. Cohen completes the proof (in a more general setting) by noting that for any space X, $X_+ \cong X \vee S^0$ stably, and hence $(X^k)_+ = (X_+)^{(k)} \cong (X \vee S^0)^{(k)} \cong \bigvee_{i=0}^k \binom{k}{i} X^{(i)}$ as Σ_k-spectra. There is a completely parallel proof of 3.2, due to M. Crabb [Cr]. In this case we have a quotient map $EU(k) \times_{\mathrm{ad}} U(k) \to F_k U$, such that composition with the projection $F_k U \to F_k U/F_{k-1}U = EU(k)^+ \wedge_{U(k)} S^{\mathrm{ad}}$ yields the obvious map induced by $U(k) \xrightarrow{\pi} S^{\mathrm{ad}}$. Again, thanks to [LMS], it would be enough to produce a map of $U(k)$-spectra $S^{\mathrm{ad}} \xrightarrow{s} U(k)$ with $\pi s = 1$. But this is easy to do explicitly, since $U(k)$, with the adjoint action, is an equivariantly framed manifold (see [Cr]).

The question now arises (G. Carlsson): does the rank filtration of $|W_\bullet \mathcal{C}|$ stably split in other cases of interest? For simplicity we focus on the case $\mathcal{C} = \mathcal{F}R$, where R has the invariant dimension property. As a first step, one might try to find $GL_k R$-complexes $C(R^k)$ analogous to $(S^1)^k$, $U(k)$. Carlsson has pointed out the following construction. Let $C(R^k)$ be the simplicial set of p-tuples of independent summands of R^k. Thus the p-simplices of $C(R^k)$ are the p-tuples $(W_1, \dots, W_p)$ of free summands of R^k such that $W_i \cap (\Sigma_{i \neq j} W_j) = 0$. Here W_i can be zero and $\bigoplus_{i=1}^p W_i$ may or may not be all of R^k. The simplicial operators are the obvious ones and may be specified by considering $C(R^k)$ as a sub-simplicial set of the bar complex for the monoid of all submodules of R^k (under addition). One can easily verify the following.

3.3 *Define $\varphi : C(R^k) \to \widetilde{B}(R^k)$ by*

$$\varphi(W_1, \dots, W_p) = (W_1 \leq W_1 \oplus W_2 \leq \dots \leq W_1 \oplus \dots \oplus W_p) \quad (\text{if } \textstyle\bigoplus W_i = R^k),$$
$$= * \quad \textit{otherwise}.$$

Then φ is a map of simplicial $GL_k R$-sets.

3.4 *Define $\eta : T(C(R^k)_p) \to (\mathcal{F}R)_p$ by*

$$(W_1, \dots, W_p) \mapsto (* \to W_1 \to W_1 \oplus W_2 \to \dots \to W_1 \oplus \dots \oplus W_p),$$

plus the obvious map on morphisms (here T denotes translation category; cf. proof of 1.2). Then η is a map of simplicial categories and so induces a map

$$|\eta| : EGL_k R \times_{GL_k R} |C(R^k)| \longrightarrow |W_\bullet^k \mathcal{F}R|.$$

Furthermore, $|\eta|$ composed with the projection $|W_\bullet^k \mathcal{F}R| \to |W_\bullet^k \mathcal{F}R|/|W_\bullet^{k-1} \mathcal{F}R| \cong EGL_k R^+ \wedge_{GL_k R} |\widetilde{B}(R^k)|$ is just $1 \wedge_{GL_k R} |\varphi|$.

Remark. Note also that $C(R^k)$ has a rank filtration:

$$C_i(R^k)_p = \{(W_1, \ldots, W_p) : \text{rank}(\bigoplus W_j) \le i\}.$$

It is easy to see that $C_i(R^k)/C_{i-1}(R^k) = GL_kR^+ \wedge_{P_i} \widetilde{A}(R^i)$ as simplicial GL_kR-sets, where:

(i) $A(R^i)$ is the complex of "split flags" — i.e. $A(R^i)_p$ consists of $(p+2)$-tuples of independent free summands spanning R^i — and $\widetilde{A}(R^i)$ is the double suspension as in Section 1. (The precise definition will be left to the reader.)

(ii) P_i is the stabilizer of R^i in R^k.

Of course the $C(R^k)$ could be defined more generally. In particular, if $\mathcal{F}R$ is replaced by Set$_*$ or $\mathbf{C}^{\text{top}}$ the corresponding complexes C_k have realizations $(S^1)^k$ and $U(k)$, respectively. Note that in these cases there is no distinction between the split flag complex A_k and the usual flag complex B_k.

In the case of $U(k)$ this amounts to eigenspace decomposition, as in Section 2. Map $(C_k)_p \times \Delta^p$ to $U(k)$ by sending $((W_0, \ldots, W_p), (a_0, \ldots, a_{p-1}))$ to the matrix defined by $A|_{W_p} = I$, $A|_{W_i} = e^{2\pi i a_i} I$ (notation as in Section 2). It is easy to check that this yields a homeomorphism $|(C_k)_\bullet| \cong U(k)$. The top stratum consists of matrices with no eigenvalue equal to one, and so is equivariantly homeomorphic to the adjoint representation (see [Cr]). Hence we obtain as promised a direct proof that the topological building $\widetilde{B}^{\text{top}}(\mathbf{C}^k)$ is S^{ad}. Furthermore the filtration described above is again just the usual one. The quotients of the filtration have the form $U(k)^+ \wedge_{U(i)\times U(k-i)} S^{\text{ad}_i}$. This is the Thom complex of the i-dimensional adjoint bundle over the Grassmannian $G_{k,i}$ of i-planes in k-space. Now consider the case of finite sets. Let $S_k = \{1, \ldots k\}$. Identify subsets $(i_1, \ldots, i_r)$ of S_k with the corresponding subspaces $\langle e_{i_1}, \ldots, e_{i_r} \rangle$ of $\mathbf{C}^k$ (here $e_1, \ldots, e_k$ is the standard basis). Then $C(S_k)$ embeds in $C(\mathbf{C}^k)$, and it is clear that the induced map $|C_\bullet(S_k)| \hookrightarrow U(k)$ is a homeomorphism onto the diagonal matrices $(S^1)^k$. Note that C_i/C_{i-1} is a wedge of $\binom{k}{i}$ i-spheres, as it should be.

The complexes $C(R^k)$ and their Borel constructions seem to be interesting objects in their own right. Unfortunately, however, there does not seem to be any general splitting theorem along the lines of the Cohen/Crabb proofs of the Kahn/Miller theorems discussed above. Indeed, as we will show in the next section, there is no splitting at all for the rank filtration when $R = \mathbf{F}_q$.

4 The case of a finite field

Let $q = p^m$, p prime, and fix a prime $\ell \neq p$. All homology and cohomology groups in this section have $\mathbf{Z}/\ell$ coefficients. Let r = order of q in $(\mathbf{Z}/\ell)^*$, $a = v_\ell(q^r - 1)$. Then (cf. [Q1]) $r = \min\{k : \ell \text{ divides } |GL_k\mathbf{F}_q|\}$ and the ℓ-Sylow subgroup of $GL_r\mathbf{F}_q$ is just the ℓ-Sylow subgroup of $\mathbf{F}_{q^r}^*$, and so is isomorphic to $\mathbf{Z}/\ell^a$.

Identifying $\mathbb{Z}/r$ with the Galois group of $\mathbb{F}_{q^r}$ over $\mathbb{F}_q$, we can form the semi-direct product $C = \mathbb{Z}/r \tilde{\times} \mathbb{Z}/\ell^a$. The inclusion $C \subset GL_r\mathbb{F}_q$ induces an isomorphism on mod ℓ homology. Let W denote the graded vector space $\widetilde{H}_*BC_+$. Thus $\dim W_i$ is 1 if $i = 2jr$ or $2jr - 1$ for some j and is zero otherwise. Of course W is also a right module over the Steenrod algebra A.

4.1 Theorem [Q1]. *$H_*BGL\mathbb{F}_q \cong S(W)$ as algebras over the Steenrod algebra. If ℓ is odd or $4|(q-1)$, $H^*BGL\mathbb{F}_q \cong S(W)$ as algebras.* ∎

The homology isomorphism above is induced by the natural map $BC \to BGL_r\mathbb{F}_q \to BGL\mathbb{F}_q$. Map BC_+ to $BGL\mathbb{F}_q^+ \times \mathbb{Z}$ by sending the disjoint basepoint to the zero component and BC to the 1-component. (Here $BGL\mathbb{F}_q^+$ is Quillen's plus construction.) Since $BGL\mathbb{F}_q^+ \times \mathbb{Z} \cong \Omega BQ\mathbb{F}_q$, this adjoints to a map $\Sigma BC_+ \xrightarrow{j} BQ\mathbb{F}_q$. Now let $V = \bigoplus_{j\geq 1} H_{2jr}CP^\infty$.

4.2 Theorem. (a) *There is an extension of Hopf algebras over the Steenrod algebra:*

$$S(V) \longrightarrow H_*BQ\mathbb{F}_q \longrightarrow S(\Sigma V_+).$$

If ℓ is odd, the extension is trivial as Hopf algebras:

$$H_*BQ\mathbb{F}_q \cong S(V) \otimes S(\Sigma V_+), \text{ and } \Sigma V_+ \subseteq \text{ primitives;}$$

(b) *$j_* : H_*\Sigma BC_+ \to H_*BQ\mathbb{F}_q$ is an isomorphism onto the primitives;*

(c) *$H^*BQ\mathbb{F}_q \cong S(\Sigma W_+)$ as algebras, where $W = \widetilde{H}_*BC_+$.*

Remark. Note (a) and (b) determine the A-algebra structure completely, since (b) shows how the exterior generators are linked by Steenrod operations to primitives in $S(V)$. $S(V)$ is of course a quotient Hopf algebra over A of H_*BU, and $S(\Sigma V_+)$ is a sub-Hopf algebra over A of H_*U.

Proof of 4.2. Since $BQ\mathbb{F}_q$ is an H-space with $\pi_1 = \mathbb{Z}$, $BQ\mathbb{F}_q \cong \widetilde{BQ\mathbb{F}_q} \times S^1$ as spaces, where $\widetilde{BQ\mathbb{F}_q}$ is the universal cover. Hence we can replace $BQ\mathbb{F}_q$ with $\widetilde{BQ\mathbb{F}_q}$, and prove the obvious analogues of (a)-(c) in which the one-dimensional generator is deleted. Out of the plethora of spectral sequences available here we choose the following. By [Q1] there is a fibre sequence

$$U \longrightarrow BGL\mathbb{F}_q^+ \longrightarrow BU \xrightarrow{\psi^q - 1} BU.$$

This is known to be a sequence of infinite loop spaces, and in particular deloops to a fibre sequence

$$BU \xrightarrow{f} \widetilde{BQ\mathbb{F}_q} \xrightarrow{q} SU \xrightarrow{h} SU.$$

We use the base-to-fibre Eilenberg-Moore spectral sequence of the fibration h. This is a second quadrant homology spectral sequence with $E^2_{p,q} = \mathrm{Cotor}^{H_*SU}_{-p,q}(\mathbb{F}_\ell, H_*SU) =$

$\mathrm{Ext}^{H^*SU}_{-p,q}(\mathsf{F}_\ell, H^*SU)$, converging to $H_{p+q}\widetilde{BQ\mathsf{F}_q}$. Now h_* induces multiplication by $q^{j-1}-1$ on π_{2j-1} and on the homology primitives. Thus on primitives in dimension $2j-1$, $h_*=0$ if $j=1 \bmod r$, and h_* is an isomorphism if $j \neq 1 \bmod r$. From this one easily computes:

$$E^2 \cong S(V) \otimes S(\Sigma V) \quad \text{as algebras.}$$

Next we show the spectral sequence collapses, by producing generators explicitly. There is a commutative diagram

$$\begin{array}{ccc} BU & \xrightarrow{f} & \widetilde{BQ\mathsf{F}_q} \\ \downarrow & & \downarrow g \\ PU & \longrightarrow & SU \\ \downarrow & & \downarrow h \\ SU & \xrightarrow{=} & SU \end{array}$$

where PU is the path space. Comparing the two spectral sequences shows the elements of V are permanent cycles and in fact correspond to $\bigotimes_{j\geq 1} H_{2jr}CP^\infty \subseteq H_*BU \xrightarrow{f_*} H_*\widetilde{BQ\mathsf{F}_q}$. Next, consider the odd dimensional generators — i.e., ΣV. These are permanent cycles precisely when they lie in the image of the edge homomorphism

$$H_*\widetilde{BQ\mathsf{F}_q} \longrightarrow \mathrm{Cotor}^{H_*SU}_{0,*}(\mathsf{F}_\ell, H_*SU) \subseteq H_*SU.$$

(Explicitly, ΣV consists of the primitives in the Hopf algebra H_*SU, in degrees of the form $2jr+1$.) But consider the composite

$$(*) \qquad \Sigma BC \xrightarrow{j} \widetilde{BQ\mathsf{F}_q} \xrightarrow{g} SU.$$

Looping this yields

$$(**) \qquad BC \longrightarrow \Omega\Sigma BC \longrightarrow BGL\mathsf{F}_q^+ \longrightarrow BU.$$

Let φ denote the composite $BC \to BU$ in $(**)$. Then by Quillen's calculations φ is injective in even dimensions. It follows easily that gj is injective in odd dimensions, and hence maps onto ΣV (since every element of $H_*\Sigma BC$ is primitive). Thus the spectral sequence collapses. When ℓ is odd there can't be any multiplicative extensions, and (a) follows, using our explicit choice of generators. Part (c) is immediate (even when $\ell=2$) since $S(V)$ and $S(\Sigma V_+)$ are self-dual Hopf algebras. To prove (b) it is enough to show j_* is injective. But this follows as in $(**)$ above, since $H_*BC \to H_*BGL\mathsf{F}_q^+$ is injective. ∎

Remark. Since $\mathrm{Tor}^{S(W)}(\mathsf{F}_\ell, \mathsf{F}_\ell)$ is isomorphic to $S(\Sigma W)$ as graded vector space, it follows that the fibre-to-base Eilenberg-Moore spectral sequence of the path-loop fibration for $BQ\mathsf{F}_q$ also collapses.

Next we compute the homology of the mod ℓ Steinberg module $St_{q,n,\ell}$. This is the $GL_n\mathbf{F}_q$ module $\widetilde{H}_{n-2}(B(\mathbf{F}_q^n), \mathbf{Z}/\ell)$. (When $n = 1$, this is to be interpreted as the trivial module.) Of course the definition makes sense for $\ell = p$ too, but then St is a projective irreducible module and so has trivial homology. Recall r is the order of q in $(\mathbf{Z}/\ell)^*$, and W_+ is the graded vector space with Poincaré series $(1+t^{2r-1})/(1-t^{2r})$. Then $\mathrm{Ext}_{S(W_+)}(\mathbf{F}_\ell, \mathbf{F}_\ell) \cong S(\Sigma W_+)$ and $\mathrm{Tor}^{S(W_+)}(\mathbf{F}_\ell, \mathbf{F}_\ell)$ is the dual algebra (exterior $\otimes$ divided power). Let x denote an exterior generator in $\mathrm{Tor}_{1,0}$ (corresponding to the zero-dimensional generator of W_+). Then we can write $\mathrm{Tor}^{S(W_+)}(\mathbf{F}_\ell, \mathbf{F}_\ell) = \mathrm{Tor}^{S(W)}(\mathbf{F}_\ell, \mathbf{F}_\ell) \oplus x \cdot \mathrm{Tor}^{S(W)}(\mathbf{F}_\ell, \mathbf{F}_\ell)$.

4.3 Theorem. *Assume ℓ is odd or $\ell = 2$ and $4|(q-1)$. Then*

$$
\begin{aligned}
H_*(GL_n\mathbf{F}_q, St_\ell) &\cong \mathrm{Tor}_{n,*}^{S(W_+)}(\mathbf{F}_\ell, \mathbf{F}_\ell) \quad (\textit{if } r = 1) \\
&\cong \mathrm{Tor}_{j,*}^{S(W)}(\mathbf{F}_\ell, \mathbf{F}_\ell) \quad (\textit{if } r > 1 \textit{ and } n = jr \textit{ or } jr+1) \\
&\cong 0 \quad (\textit{otherwise}).
\end{aligned}
$$

Proof. The reduced chain complex of $\widetilde{B}_n$ has the form

$$0 \longleftarrow C_0 \longleftarrow C_1 \longleftarrow \cdots \longleftarrow C_n \longleftarrow 0$$

where $C_0 = 0$, $C_1 = \mathbf{F}_\ell$ and, for $p > 1$, $C_p = \bigotimes_{|I|=p} 1_{P_I}^G$. Here $G = GL_n\mathbf{F}_q$, $I = (i_1, \ldots, i_p)$ is a partition of n, and P_I is the corresponding parabolic. By a standard argument we get a first quadrant homology spectral sequence with $E^1_{p,q} = \bigoplus_{|I|=p} H_q(G, C_p)$, converging to $H_{p+q-n}(G, St)$. Now $H_*(G, 1_{P_I}^G) = H_*(P_I, \mathbf{F}_\ell)$ by Shapiro's Lemma, and the projection of P_I onto its Levi factor $G_I = G_{i_1} \times \ldots \times G_{i_p}$ is an isomorphism on mod ℓ homology since the kernel is a p-group and $\ell \neq p$. Hence $E^1_{p,q} = \bigoplus_{|I|=p} H_q G_I$. Now consider the case $r = 1$. $S(W_+)$ has a second grading, obtained by assigning grade 1 to W_+. Now the bar complex of any graded algebra splits up into homogeneous pieces; let $D_{*,n}$ denote the nth graded piece (so $*$ denotes the homological degree). Then in our case (a) $D_{k,k} = 0$ if $k > n$, (b) $D_{*,n}$ is acyclic below degree n, and (c) the subgroup of cycles in $D_{n,n}$ is precisely $\mathrm{Tor}_n^{S(W_+)}(\mathbf{F}_\ell, \mathbf{F}_\ell)$. Here (a) is obvious and (b), (c) are easily seen as follows. The cohomology of the dual complex is $\mathrm{Ext}_{S(W_+)}(\mathbf{F}_\ell, \mathbf{F}_\ell) \cong S(\Sigma W_+)$. Since the elements of ΣW_+ here have both grade 1 and homological degree 1, it is clear that Ext^n is concentrated in grade n. Hence the same is true of Tor_n, and our assertion follows. Returning to the spectral sequence, we have $H_* G_i = S^i(W_+)$ (by Quillen [Q1]; here we use our assumption on ℓ) and hence $E^1_{p,*} = \bigoplus_{\ell(I)=p} S^I(W_+)$, where $S^I = S^{i_1} \otimes \ldots \otimes S^{i_p}$. But it is clear on inspection that d^1 is exactly the differential in the (reduced) bar complex for $S(W_+)$. Thus the E^1 term is just the n-th graded piece of the bar complex, and E^2 is just $\mathrm{Tor}_n^{S(W_+)}$, concentrated on the line $p = n$. Hence the spectral sequence collapses at E^2 and the theorem follows for $r = 1$.

The proof for $r > 1$ is similar. In this case we grade $S(W_+)$ by assigning grade 1 to the zero-dimensional part of W_+ and grade r to W. Again the bar complex splits into homogeneous pieces $D_{*,n}$, but now Ext, and hence Tor, are concentrated in grades congruent to zero or one mod r. It follows that $D_{*,n}$ is acyclic if $n \neq 0, 1 \bmod r$. If $n = jr$, then $D_{k,n}$ is acyclic for $k < j$, and the group of cycles in $D_{j,n}$ is $\mathrm{Tor}_j^{S(W)}(\mathsf{F}_\ell, \mathsf{F}_\ell)$. If $n = jr + 1$, $D_{k,n}$ is acyclic for $k < j + 1$, and the group of cycles in $D_{j+1,n}$ is $x \cdot \mathrm{Tor}_j^{S(W)}(\mathsf{F}_\ell, \mathsf{F}_\ell)$. Now the the spectral sequence has E^1-term as before: $E^1_{p,*} = \bigoplus_{\ell(I)=p} N_I$, where $N_I = H_*GL_{i_1} \otimes \ldots \otimes H_*GL_{i_p}$. Let b_0 denote a generator of H_0BGL_1, and set $i_s = j_s r + \varepsilon_s$, $0 \leq \varepsilon_s < r$. Then $H_*BGL_{i_s} = [S^{j_s}(H_*BGL_r)] \cdot b_0^{\varepsilon_s}$. Hence $E^1_{p,*}$ (p fixed) can be identified with $D_{p,n}$. The theorem now follows as in the case $r = 1$. ∎

4.4 **Corollary.** *The rank filtration splits on mod ℓ homology.*

Proof. In other words, the spectral sequence associated to the rank filtration collapses at E^1. Since $E^1 = \mathrm{Tor}^{S(W_+)}(\mathsf{F}_\ell, \mathsf{F}_\ell) = H_*BQ\mathsf{F}_q$ as graded vector spaces, this is immediate. ∎

To conclude we note that there is no splitting over the Steenrod algebra. Consider for example the case $r = 1$. Then $F_1(H_*BQ\mathsf{F}_q) = j_*H_*\Sigma(B\mathbb{Z}/\ell^a_+)$, where $j: \Sigma BC_+ \to BQ\mathsf{F}_q$, under j_*, and we have seen that this is precisely the subspace of primitives. Suppose this was a direct summand of $H_*BQ\mathsf{F}_q$. Then the indecomposables of $H^*BQ\mathsf{F}_q$ also split off. But this implies all pth powers are zero, contradicting Theorem 4.2(c).

References

[B-S] A. Borel and J.P. Serre: Corners and arithmetic groups. *Comment. Math. Helv.* 48 (1974), 244–297.

[Co] R. Cohen: Stable proofs of stable splittings. *Math. Proc. Cambridge Philos. Soc.* 88 (1980), 149–151.

[Cr] M. Crabb: On the stable splitting of $U(n)$ and $\Omega U(n)$. *Lecture Notes in Math. 1298, pp. 35–53, Springer*, 1987.

[G] D. Grayson: Higher algebraic K-theory: II (after Daniel Quillen). *Lecture Notes in Math. 551, pp. 217–240, Springer*, 1976.

[L-M-S] L. Lewis, Jr., J.P. May and M. Steinberger: Equivariant Stable Homotopy Theory. Lecture Notes in Math. 1213, Springer, 1986.

[Mil] H. Miller: Stable splittings of Stiefel manifolds. *Topology* 24 (1985), 411–419.

[Mit] S. Mitchell: Quillen's theorem on buildings and the loops on a symmetric space. To appear in *Enseign. Math.*

[P] S.B. Priddy: On $\Omega^\infty S^0$ and the infinite symmetric group. *Proc. Sympos. Pure Math.* 22 (1971), 217–220.

[Q1] D. Quillen: On the cohomology and K-theory of the general linear groups over a finite field. *Ann. of Math.* 96 (1972), 552–586.

[Q2] D. Quillen: Higher algebraic K-theory. I. *Lecture Notes in Math. 341, pp. 85–147, Springer*, 1973.

[Q3] D. Quillen: Finite generation of the groups K_i of rings of algebraic integers. *Lecture Notes in Math. 341, pp. 179–198, Springer*, 1973.

[W] F. Waldhausen: Algebraic K-theory of topological spaces. I. *Proc. Sympos. Pure Math* 32 (1978), 35–60.

Obstruction theory and the strict associativity of Morava K-theories

Alan Robinson

Introduction

Let E be a spectrum, in the sense of stable homotopy theory. According to a standard definition, E is a ring spectrum if there is an associative ring structure in the cohomology theory represented by E. (We shall not be concerned with commutativity conditions.) Equivalently, one may say that E is a ring spectrum if it has a multiplication map $\mu: E \wedge E \to E$ and a unit map $\eta: S \to E$ such that the following diagrams commute up to homotopy:

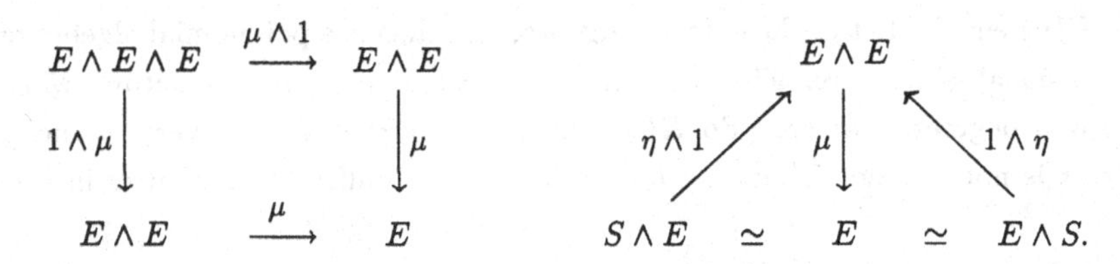

This definition is adequate for many purposes. For others, such as the construction of Massey products and of the derived category of a ring spectrum, one needs higher associativity conditions like those which Stasheff introduced into the theory of H-spaces. If E has this extra structure, it is called an A_∞ ring spectrum. There is an analogue of a theorem of Adams about H-spaces: a spectrum admits an A_∞ ring structure if and only if it is homotopy equivalent to a strictly associative ring spectrum. Indeed, in a category of spectra with strictly associative smash products (such as that of Elmendorf [4]), one can easily adapt the proof of the theorem in [3, §8.2] to the case of spectra.

In this paper, we develop the theory of obstructions to refining a ring structure on a spectrum E to an A_∞ ring structure (or, equivalently, obstructions to finding a strictly associative ring spectrum which is homotopy equivalent to E). We have to assume that the E-cohomology of the smash powers $E^{(r)}$ of E is given by perfect Künneth and duality theorems:

$$E^*(E^{(r)}) \approx \mathrm{Hom}((E_*E)^{\otimes r}, E_*),$$

where the homomorphisms and tensor product are taken over the coefficient ring E_*. This granted, we show that the obstructions lie in the Hochschild cohomology of the dual Steenrod algebra (or ring of co-operations) E_*E. Since E_*E is graded, its Hochschild cohomology is bigraded. The obstructions to existence of an A_∞ structure all lie in odd total degrees. There are also obstructions to uniqueness, which lie in even total degrees. There is even a spectral sequence which begins with the cohomology of E_*E and converges to the homotopy groups of the space of A_∞ structures on E.

As an application of the theory, we consider the case in which E is Morava's nth K-theory $K(n)$ for some odd prime p and some integer $n \geq 1$. The ring structure on $K(n)$ is usually obtained by regarding it as a unitary bordism theory with singularities in the sense of Sullivan [7, 8, 12]. There are technical problems with product singularities which are not very easy to resolve. For this reason it would not be easy to extend this method to prove that $K(n)$ has an A_∞ structure. On the other hand, our obstruction theory works well. The structure of the dual Steenrod algebra is known from work of Morava (unpublished) and Yagita [16]:

$$K(n)_*K(n) \approx \frac{K(n)_*[t_1, t_2, \ldots]}{(v_n t_i^{p^n} - v_n^{p^i} t_i)} \otimes \Lambda(\tau_0, \tau_1, \ldots, \tau_{n-1}).$$

The first factor on the right turns out to be acyclic, so the Hochschild cohomology of $K(n)_*K(n)$ equals that of the exterior algebra, and thus is a polynomial algebra on even-dimensional generators. Therefore the obstructions to an A_∞ structure, which all lie in odd degrees, are zero. So $K(n)$ has an A_∞ structure for every n and p. However, it is not unique. There are uncountably many different structures in each case.

This result can be paraphrased as follows. There are uncountably many strictly associative ring spectra homotopy equivalent to any given $K(n)$, such that the standard equivalence from one to any other does not admit the structure of an A_∞ homomorphism.

1 The obstruction theory of A_∞ ring structures

1.1 The basic definition is the following (cf. [10]). Let K_n be the Stasheff cell homeomorphic to D^{n-2}, and $\partial_j^{r,s}: K_r \times K_s \to K_{r+s-1}$, $(1 \leq j \leq r)$, the face inclusions. An A_n *ring structure* on a spectrum E is defined to be a family of maps $\mu_m: K_m^+ \wedge E^{(m)} \to E$, $(2 \leq m \leq n)$ such that

$$\mu_{r+s-1}\partial_j^{r,s} = \mu_r \cdot (1^{(j-1)} \wedge \mu_s \wedge 1^{(r-j)}) \tag{1.2}$$

as maps from $(K_r \times K_s)^+ \wedge E^{(r+s-1)} \approx K_r^+ \wedge E^{(j-1)} \wedge (K_s^+ \wedge E^{(s)}) \wedge E^{(r-j)}$ to E, and such that the multiplication $\mu_2: E^{(2)} \approx K_2^+ \wedge E^{(2)} \to E$ has a two-sided homotopy unit. (As usual, $E^{(k)}$ here denotes the kth smash power of E.)

1.3 Suppose that E is a ring spectrum in the traditional sense. Thus E has a multiplication $\mu = \mu_2\colon E^{(2)} \to E$ for which there exist a homotopy unit and an association homotopy $\mu(\mu \wedge 1) \simeq \mu(1 \wedge \mu)$. This is exactly the same as saying that E has an A_2 ring structure which can be extended (perhaps in more than one way) to an A_3 ring structure.

We develop an obstruction theory for extending a given A_{n-1} structure to an A_n structure. The existence of such an obstruction theory has been known in principle since the original papers of Stasheff [13], at least in relation to the theory of H-spaces. Our contribution is to show that for ring spectra the obstruction theory frequently reduces to a feasible calculation in Hochschild cohomology.

1.4 **Assumption.** We assume that the ring spectrum E satisfies the following condition, which holds in many cases of interest. The E-cohomology of each smash power $E^{(k)}$ can be calculated from $E_*(E)$ by perfect Künneth and duality theorems:

$$E^*(E^{(k)}) \approx \mathrm{Hom}_{E_*}((E_*E)^{\otimes k}, E_*), \quad k \geq 1.$$

This is a moderately severe restriction: see [1, p.284]. In order to relieve the notation, we shall write R for the coefficient ring E_*, and Λ for the dual Steenrod algebra E_*E. The R-algebra structure on Λ is given by the "left unit" $\eta_L\colon R \to \Lambda$. Thus

$$E^*(E^{(k)}) \approx \mathrm{Hom}_R(\Lambda^{\otimes k}, R),$$

where the tensor product is over R.

1.5 **The obstruction cochain.** Suppose that we already have an A_{n-1} structure $\{\mu_r\}_{r\leq n-1}$ on E, where $n \geq 3$. To extend it to an A_n structure, we need a map $\mu_n\colon K_n^+ \wedge E^{(n)} \to E$. The compatibility conditions (1.2) define μ_n precisely on $\partial K_n^+ \wedge E^{(n)}$, which is homeomorphic to $S^{n-3} \wedge E^{(n)}$. Since $K_n \approx D^{n-2}$, the obstruction to extending μ_n over $K_n^+ \wedge E^{(n)}$ is an element $c_n \in E^0(\partial K_n^+ \wedge E^{(n)}) \approx E^{3-n}(E^{(n)})$. By the hypothesis of 1.4, we therefore have

$$\begin{aligned} c_n \in E^{3-n}(E^{(n)}) &\approx \mathrm{Hom}_R^{3-n}(E_*(E^{(n)}), R) \\ &\approx \mathrm{Hom}_R^{3-n}(\Lambda^{\otimes n}, R) \\ &\approx C^{n,3-n}(\Lambda; R), \end{aligned}$$

where $C^{n,*}(\Lambda; R)$ is the group of Hochschild n-cochains of the graded R-algebra Λ with values in R. Thus we have proved the following.

1.6 **Lemma.** *The given A_{n-1} structure on E extends to an A_n structure if and only if the obstruction cochain $c_n \in C^{n,3-n}(\Lambda; R)$ is zero.* ∎

We shall show later that for $n \geq 4$ the cochain c_n is a Hochschild cocycle.

1.7 Enumeration by difference classes. In the situation of 1.5, suppose that the given A_{n-1} structure extends in at least one way to an A_n structure. Then, since $(K_n, \partial K_n) \approx (D^{n-2}, S^{n-3})$, the homotopy classes of possible maps μ_n with the required restriction to the boundary are enumerated by difference classes in the group

$$\begin{aligned}[S^{n-2} \wedge E^{(n)}, E] &\approx E^{2-n}(E^{(n)}) \\ &\approx \mathrm{Hom}_R^{2-n}(\Lambda^{\otimes n}, R) \\ &\approx C^{n,2-n}(\Lambda; R).\end{aligned}$$

1.8 Calculation of the coboundary map

Proposition. *Suppose that $n \geq 4$. If we alter the A_{n-1} structure on E by a difference class $a \in C^{n-1,3-n}(\Lambda; R)$, leaving the A_{n-2} structure unchanged, then the obstruction $c_n \in C^{n,3-n}(\Lambda; R)$ to an A_n structure is changed by an amount equal to the Hochschild coboundary δa of a.*

Proof. The old and new obstruction cochains are represented by maps $\partial K_n^+ \wedge E^{(n)} \to E$ which are identical on the $(n-4)$-skeleton of the cell complex $\partial K_n \approx S^{n-3}$. We must calculate the difference element between them on each top-dimensional face $\partial_j^{r,s}(K_r \times K_s)$ of ∂K_n, $(r + s = n + 1,\ 1 \leq j \leq r)$. By the relations (1.2), the restriction to this face is determined by the maps μ_r and μ_s. So, if $r, s < n-1$, there is no difference between the old and new maps on this face. There remain the faces $\partial_j^{n-1,2}$ $(1 \leq j \leq n-1)$ and $\partial_j^{2,n-1}$, $(j = 1, 2)$.

On the face $\partial_j^{n-1,2}$, (1.2) shows that the difference class is represented by the map

$$E^{(n)} \approx E^{(j-1)} \wedge E^{(2)} \wedge E^{(n-j-1)} \xrightarrow{1 \wedge \mu_2 \wedge 1} E^{(j-1)} \wedge E \wedge E^{(n-j-1)} \xrightarrow{\alpha} S^{3-n}E$$

where α represents the difference class between the new and old maps μ_{n-1}. This corresponds under the duality isomorphism of 1.4 to the n-cochain

$$\Lambda^{\otimes n} \approx \Lambda^{\otimes(j-1)} \otimes \Lambda^{\otimes 2} \otimes \Lambda^{\otimes(n-j-1)} \xrightarrow{1 \wedge \mu \wedge 1} \Lambda^{\otimes(n-1)} \xrightarrow{a} R. \qquad (1.9)$$

On the face $\partial_1^{2,n-1}$, the difference is represented by the map

$$E^{(n)} \approx E \wedge E^{(n-1)} \xrightarrow{1 \wedge \alpha} E \wedge S^{3-n}E \xrightarrow{\mu_2} S^{3-n}E. \qquad (1.10)$$

The corresponding element of $C^{n,3-n}(\Lambda; R) \approx \mathrm{Hom}_R^{3-n}(\Lambda^{\otimes n}, R)$ is the composite of the E-homology homomorphism $\Lambda^{\otimes n} \to \Lambda$ induced by (1.10) and the counit $\varepsilon: \Lambda \to R$. Let $\hat{a}: \Lambda^{\otimes(n-1)} \to \Lambda$ be the E-homology homomorphism corresponding to the given difference class $a \in C^{n-1,3-n}(\Lambda; R)$. Thus $\varepsilon \cdot \hat{a} = a$. Since (1.10) induces $\mu \cdot (1 \otimes \hat{a})$ on homology, the commutative diagram

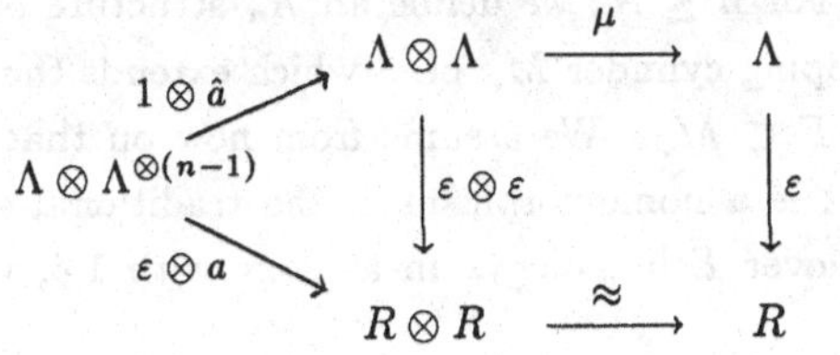

$$\begin{array}{ccccc} & \nearrow^{1\otimes \hat{a}} & \Lambda\otimes\Lambda & \xrightarrow{\mu} & \Lambda \\ \Lambda\otimes\Lambda^{\otimes(n-1)} & & \downarrow{\scriptstyle \varepsilon\otimes\varepsilon} & & \downarrow{\scriptstyle \varepsilon} \\ & \searrow_{\varepsilon\otimes a} & R\otimes R & \xrightarrow{\approx} & R \end{array}$$

shows that the cochain $\varepsilon\cdot\mu\cdot(1\otimes\hat{a})$ representing the difference class on this face is equal to $\varepsilon\otimes a$. Similarly, on the face $\partial_2^{2,n-1}$ the difference term is $a\otimes\varepsilon$.

The change in the obstruction is the sum of the difference terms over all the faces. The result of summing the homomorphisms (1.9), $\varepsilon\otimes a$ and $a\otimes\varepsilon$ with the appropriate signs is precisely the Hochschild coboundary δa, given by

$$\delta a(\lambda_1\otimes\cdots\otimes\lambda_n)=\varepsilon(\lambda_1)a(\lambda_2\otimes\cdots\otimes\lambda_n)+\sum_{i=1}^{n-1}(-1)^i a(\lambda_1\otimes\cdots\otimes\lambda_i\lambda_{i+1}\otimes\cdots\otimes\lambda_n)$$
$$+(-1)^n a(\lambda_1\otimes\cdots\otimes\lambda_{n-1})\cdot\varepsilon(\lambda_n).$$

■

1.11 Theorem. *Let E be a ring spectrum satisfying the condition 1.4. Suppose E has an A_{n-1} structure, where $n\geq 4$. Then the obstruction cochain c_n of 1.5 is a cocycle, and the underlying A_{n-2} structure on E can be refined to an A_n structure if and only if this cocycle represents zero in the Hochschild cohomology group $HH^{n,3-n}(\Lambda;R)$ of the dual Steenrod algebra $\Lambda=E_*E$ with coefficients in $R=E_*$.*

Proof. Consider the problem of defining a map $K_{n+1}^+\wedge E^{(n+1)}\to E$ satisfying the relations (1.2) for $\mu_{n+1}|\partial K_{n+1}$. The maps μ_r, $2\leq r\leq n-1$, specify it on the $(n-3)$-skeleton of $\partial K_{n+1}\approx S^{n-2}$, and there is an obstruction to extending over each top-dimensional cell. The sum of these obstructions is zero, because ∂K_{n+1} is a cycle. By the calculation used in the proof of the last proposition, with obstructions in place of difference classes, this sum is exactly the Hochschild coboundary δc_n, provided $n+1\geq 4$. Therefore $\delta c_n=0$, and c_n is a cocycle.

According to 1.8, we can alter the cocycle c_n by any coboundary, if $n\geq 4$, simply by changing the A_{n-1} structure whilst leaving the A_{n-2} structure unaltered. Therefore the A_{n-2} structure extends to an A_n structure if and only if c_n is a coboundary, that is, if it represents zero in $HH^{n,3-n}(\Lambda;R)$. ■

For $n\leq 3$, the analysis is not so simple because the fundamental nonlinearity of the relations (1.2) intervenes.

1.12 A_n structures on homomorphisms. Let E and F be A_N ring spectra, and $\varphi: E \to F$ a map. For $n \leq N$, we define an A_n structure on φ to be an A_n ring structure on the mapping cylinder M_φ of φ which extends the given A_n structures on the subspectra $E, F \subset M_\varphi$. We assume from now on that φ has at least an A_2 structure, so that it is a homomorphism in the traditional sense, and F-homology is a module theory over E-homology. In analogy with 1.4, we assume from now on the relation

$$\begin{aligned} F^*(E^{(k)}) &\approx \mathrm{Hom}_{E_*}((E_*E)^{\otimes k}, F_*), \quad k \geq 1 \\ &\approx \mathrm{Hom}_R(\Lambda^{\otimes k}, F_*). \end{aligned} \tag{1.13}$$

Suppose we wish to extend an A_{n-1} structure on φ to an A_n structure. Since $M_\varphi = (I^+ \wedge E) \cup_\varphi F$, this amounts to constructing a map of $K_n^+ \wedge (I^+ \wedge E)^{(n)}$ into M_φ having given restriction to ∂K_n and to ∂I. Homotopically, this is equivalent to constructing a map $(K_n \times I)^+ \wedge E^{(n)} \to F$ with given restriction to $\partial(K_n \times I)^+$. By the methods of 1.5–1.11, we obtain the following result.

1.14 Theorem. *Suppose $\varphi: E \to F$ is a map of ring spectra as in 1.12 and (1.13), and $n \geq 3$. Then the following hold.*

(i) *The obstruction to refining a given A_{n-1} structure on φ to an A_n structure is a cocycle $b_n \in C^{n,2-n}(\Lambda; F_*)$.*

(ii) *The effect on b_n of altering the A_{n-1} structure, whilst the A_{n-2} structure remains the same, is to add to it a general coboundary.*

(iii) *The underlying A_{n-2} structure on φ extends to an A_n structure if and only if b_n represents zero in the Hochschild cohomology group $HH^{n,2-n}(\Lambda; F_*)$.* ∎

1.15 Remarks. (i) An interesting case is when φ is Quillen's idempotent homomorphism ξ from the A_∞ ring spectrum $MU_{(p)}$ to itself [9]. Does ξ admit an A_∞ structure? If it can be shown that the obstruction classes vanish in this case, then the Brown-Peterson spectrum BP admits an A_∞ multiplication.

(ii) The obstruction theory can be further developed into a half-plane spectral sequence for the homotopy groups of the space of A_∞ structures on the ring spectrum E. This arises from the homotopy exact couple of a tower of fibrations.

$$\mathcal{A}_\infty(E) = \lim_n \mathcal{A}_n(E) \to \cdots \to \mathcal{A}_n(E) \to \mathcal{A}_{n-1}(E) \to \cdots \to \mathcal{A}_2(E)$$

where $\mathcal{A}_n(E)$ is the space of A_n structures on E. The spectral sequence has the general form

$$E^2_{-s,-t} \approx HH^{s,t+2}(\Lambda; R) \Longrightarrow \pi_{-s-t}\mathcal{A}_\infty(E),$$

but there are problems at the edge, as the homotopy couple runs out of exactness at π_0. We leave the details to the reader.

2 Strictly associative multiplications on Morava K-theories

2.1 For every prime number p and every integer $n \geq 1$, Morava introduced a homology theory $K(n)_*(\bullet)$ having coefficient ring $\pi_* K(n) \approx F_p[v_n, v_n^{-1}]$ where $\dim v_n = 2(p^n - 1)$. When $n = 1$, this theory is a summand in complex K-theory modulo p. It is known that the spectrum $K(n)$ is a ring spectrum in the traditional sense: in every case, there is a homotopy associative multiplication $\mu: K(n) \wedge K(n) \to K(n)$. (When p is an odd prime, this multiplication is unique and is homotopy commutative as well.) Thus we are in a position to apply the theory developed in Section 1.

When p is odd, the dual Steenrod algebra for $K(n)$ takes the form

$$K(n)_* K(n) \approx \frac{K(n)_*[t_1, t_2, \ldots]}{(v_n t_i^{p^n} - v_n^{p^i} t_i)} \otimes \Lambda_{K(n)_*}(\tau_0, \tau_1, \ldots, \tau_{n-1}).$$

(see [16,17]) where $\dim t_i = 2(p^i - 1)$, $\dim \tau_i = 2p^i - 1$, and Λ denotes an exterior algebra. As in Section 1, we shall denote the coefficient ring $K(n)_*$ by R.

2.2 **Theorem.** *When p is odd, the Hochschild cohomology of $K(n)_* K(n)$ is a polynomial algebra*

$$HH^{**}(K(n)_* K(n); R) \approx R[\alpha_0, \alpha_2, \cdots, \alpha_{n-1}],$$

where $\alpha_i \in HH^{1, 2p^i - 1}$. In particular, the cohomology in odd total degrees is zero.

Proof. For each $i \geq 1$, let Γ_i be the subalgebra $R[t_i]/(v_n t_i^{p^n} - v_n^{p^i} t_i)$ of $K(n)_* K(n)$. As R is a graded field, the splitting of the denominator into relatively prime factors t_i and $t_i^{p^n - 1} - v_n^{p^i - 1}$ leads to a direct product decomposition

$$\Gamma_i \approx R \times R[t_i]/(t_i^{p^n - 1} - v_n^{p^i - 1}).$$

The augmentation homomorphism $K(n)_* K(n) \to R$ takes t_i to zero [1, page 112], so its restriction to Γ_i is the projection on the first factor $\Gamma_i \to R$. By [6, Chap. X Theorem 6.2], $HH^{**}(\Gamma_i; R) \approx HH^{**}(R; R) \approx R$.

It is well known that the cohomology of the exterior algebra $\Lambda_R(\tau_j)$ is a polynomial algebra $R[\alpha_j]$, where α_j has bidegree $(1, \deg \tau_j) = (1, 2p^i - 1)$.

Now there is a tensor product decomposition

$$K(n)_* K(n) \approx \bigotimes_{i=1}^{\infty} \Gamma_i \otimes \bigotimes_{j=1}^{n-1} \Lambda_R(\tau_j).$$

The cohomology of any finite tensor subproduct can be calculated from the Künneth theorem [6, Chap. X Theorem 7.4]. The cohomologies of these finite stages form an inverse system which is eventually constant, because the factors Γ_i have trivial cohomology. Therefore the cohomology of the colimit $K(n)_* K(n)$ is isomorphic to this limit, which is $K(n)_*[\alpha_1, \alpha_2, \cdots, \alpha_{n-1}]$, as claimed. ∎

2.3 Theorem. *For every odd prime p and every integer $n \geq 1$, the periodic Morava K-theory $K(n)$ and the connective Morava K-theory $k(n)$ admit A_∞ multiplicative structures.*

Proof. For elementary reasons, $K(n)$ satisfies the hypothesis of 1.4 (see [15, §8] for instance). By results of Mironov [7] and of Shimada and Yagita [12], $K(n)$ admits an A_2 multiplication which can be refined to an A_3 multiplication. (Indeed, as we noted in 1.3, this is just the homotopy associativity condition.)

We can now use recursion to prove that $K(n)$ has an A_∞ structure. For each $m \geq 4$ in turn, we apply Theorem 1.11 to deduce that the A_{m-2} structure on $K(n)$ can be refined to an A_m structure. The relevant obstruction class lies in the odd-dimensional part of $HH^{**}(K(n)_*K(n); R)$, and is therefore zero by Theorem 2.2.

The corresponding result for $k(n)$ follows by lifting to the connective cover, in the usual way. ∎

We now turn to the enumeration of the possible A_m structures on $K(n)$. We regard two A_m structures as equivalent if the identity map $1_{K(n)}$ between them admits the structure of an A_m homomorphism. According to Theorem 1.14, there are obstructions to equivalence lying in $HH^{k,2-k}(K(n)_*K(n); R)$: these are the cohomology classes described in 1.7. To find all the A_m structures, we calculate these cohomology groups (which are all vector spaces over F_p) and show that all difference classes can be realized. The reader may like to think of the following proposition in terms of the spectral sequence introduced informally in 1.15(ii): to calculate π_0 of the space of A_∞ structures, we need to (i) find the terms $E^2_{-s,-t}$ with $s+t=0$, (ii) show there are no non-zero boundaries, and (iii) show that all the elements are infinite cycles. As before, p is taken to be an odd prime.

2.4 Proposition. (i) *The dimension over the field F_p of the group $HH^{k,2-k}(K(n)_*K(n); R)$ is equal to the number of sequences $(i_1, i_2, \ldots, i_k)$ of integers such that $0 \leq i_1 \leq i_2 \leq \cdots \leq i_k \leq n-1$ and $p^{i_1} + p^{i_2} + \cdots + p^{i_k} \equiv 1 \bmod (p^n - 1)$.*

(ii) *Let $k < m \leq \infty$. If two A_m structures on $K(n)$ have equivalent underlying A_{k-1} structures, then their A_k structures are separated by a well-defined difference class in $HH^{k,2-k}(K(n)_*K(n); R)$.*

(iii) *Every element of $HH^{k,2-k}(K(n)_*K(n); R)$ is a difference class in this sense between any given A_m structure and some other A_m structure.*

Proof. (i) We know from Theorem 2.2 that a basis for the group $HH^{k,*}(K(n)_*K(n); R)$ consists of the elements $v_n^r \alpha_{i_1} \alpha_{i_2} \ldots \alpha_{i_k}$, $0 \leq i_j \leq n-1$, $r \in \mathbf{Z}$. The congruence is the condition that this element should have bidegree $(k, 2-k)$ for some choice of r.

(ii) When an A_{k-1} structure has been chosen for the identity map $1_{K(n)}$ between the two structures, the obstruction cocycle is given by 1.14(i) (or by 1.7). One might expect its cohomology class to depend upon the choice, but the vanishing of the groups $HH^{i,1-i}(K(n)_*K(n); R)$ implies that the A_{k-2} structure on $1_{K(n)}$ is unique, and changing the A_{k-1} structure on $1_{K(n)}$ only alters the cocycle by a coboundary, according to 1.14(ii). Therefore the difference class in $HH^{k,2-k}(K(n)_*K(n); R)$ is well-defined.

(iii) We represent the desired difference class by a Hochschild cocycle ζ. By 1.7, we can take the given A_m structure and alter its underlying A_k structure by ζ. As ζ is a cocycle, Proposition 1.8 guarantee that the new A_k structure extends to an A_{k+1} structure. Then we can use Theorem 1.11 to extend the structure stage by stage to an A_m structure, because of the vanishing of the groups $HH^{i,3-i}(K(n)_*K(n); R)$. ∎

2.5 Theorem. *Let p be an odd prime, and $n \geq 1$ an integer. Then*
(i) *$K(n)$ has exactly one A_{p-1} structure which can be further extended;*
(ii) *$K(n)$ has exactly p distinct A_p structures which can be further extended;*
(iii) *$K(n)$ has uncountably many distinct A_∞ structures.*

Proof. The congruence in 2.4(i) has no solution when $k \leq p-1$. So 1.14 shows that there is no obstruction to constructing an A_{p-1} equivalence between any two A_m structures when $m \geq p$. This proves (i).

When $k = p$ the congruence has the unique solution $i_1 = i_2 = \cdots = i_k = n-1$. Proposition 2.4 therefore shows that there are exactly p distinct A_p structures which extend to A_{p+1} structures (and all of these extend to A_∞ structures). Hence (ii).

It is easy to see, using 2.4(i), that there are infinitely many values of k for which the group $HH^{k,2-k}(K(n)_*K(n); R)$ is non-zero. (There are even infinitely many values of k for which $i_1 = i_2 = \cdots = i_k = 0$ is a solution of the congruence.) At each of these, there is a finite non-trivial choice of A_k structures. Together, these choices generate an uncountable set of A_∞ structures topologized as a Cantor set. ∎

The author conjectures that the above result holds for $p = 2$ also. In fact, 2.5(i) and (ii) for $p = 2$ are equivalent to the statement that $K(n)$ has exactly two homotopy associative multiplications. This is a known result: it follows from work of Araki and Toda [2] for $n = 1$, and from Würgler [14] for general n.

2.6 Corollary. *If p is an odd prime, then there are uncountably many inequivalent strictly associative models for each $K(n)$.*

Proof. This is the translation of 2.5(iii) which one obtains by converting the A_∞ structures into strictly associative multiplications on homotopically equivalent spectra. "Inequivalent" here means that the standard homotopy equivalence between two models admits no A_∞ homomorphism structure. ∎

Finally, we record one consequence of the existence of an A_∞ product on $k(n)$.

2.7 Corollary. *For each odd prime p and each integer $n \geq 1$, there is a universal coefficient theorem for $k(n)$ in the form of an exact sequence*

$$0 \to \mathrm{Ext}^{1,m-1}_{k(n)_*}(k(n)_*(X), k(n)_*) \to k(n)^m(X) \to \mathrm{Hom}^m_{k(n)_*}(k(n)_*(X), k(n)_*) \to 0.$$

Proof. After Theorem 2.3, this follows immediately from [11, page 257, Corollary]. ∎

The corresponding Künneth theorem for connective Morava K-theory $k(n)$ was proved by Lellmann [5], and independently by A.G.K. Modawi and by D. Whitgift.

References

[1] J. F. Adams: Stable homotopy and generalised homology. University of Chicago Press, Chicago, 1974.

[2] S. Araki and H. Toda: Multiplicative structures in mod q cohomology theories. I. *Osaka J. Math.* 2 (1965), 71–115; II. ibid. 3 (1966), 81–120.

[3] J.M. Boardman: Homotopy structures and the language of trees. *Proc. Sympos. Pure Math.* 22 (1971), 37–58.

[4] A. Elmendorf: The Grassmannian geometry of spectra. *J. Pure Appl. Alg.* 54 (1988), 37–94.

[5] W. Lellmann: Connected Morava K-theories. *Math. Z.* 179 (1982), 387–399.

[6] S. MacLane: Homology. Springer, Berlin-Göttingen-Heidelberg, 1963.

[7] O.K. Mironov: Multiplications in cobordism theories with singularities, and Steenrod-tom Dieck operations. *Izv. Akad. Nauk SSSR Ser. Mat.* 42 (1978), 789–806 = *Math. of USSR - Izvestiya* 13 (1979), 89–106.

[8] J. Morava: Product for the odd-primary bordism of manifolds with singularities. *Topology* 18 (1979), 177–186.

[9] D.G. Quillen: On the formal group laws of unoriented and complex cobordism theory. *Bull. Amer. Math. Soc.* 75 (1969), 1293–1298.

[10] A. Robinson: Derived tensor products in stable homotopy theory. *Topology* 22 (1983), 1–18.

[11] A. Robinson: Spectra of derived module homomorphisms. *Math. Proc. Cambridge Philos. Soc.* 101 (1987), 249–257.

[12] N. Shimada and N. Yagita: Multiplications in the complex bordism theory with singularities. *Publ. Res. Inst. Math. Sci., Kyoto University* 12 (1976), 259–293.

[13] J.D. Stasheff: Homotopy associativity of H-spaces. I. *Trans. Amer. Math. Soc.* 108 (1963), 275–292; II. ibid 293–312.

[14] U. Würgler: Commutative ring spectra of characteristic 2. *Comment. Math. Helv.* 61 (1986), 33–45.

[15] U. Würgler: On products in a family of cohomology theories associated to the invariant prime ideals of $\pi_*(BP)$. *Comment. Math. Helv.* 52 (1977), 457–481.

[16] N. Yagita: A topological note on the Adams spectral sequence based on Morava's K-theory. *Proc. Amer. Math. Soc.* 72 (1978), 613–617.

[17] N. Yagita: On the Steenrod algebra of Morava K-theory. *J. London Math. Soc.* (2), 22 (1980), 423–438.

Loop groups and harmonic maps

Graeme Segal

Introduction

The main purpose of this note is to give a short account of the work of Karen Uhlenbeck [10] on the harmonic maps from the Riemann sphere to a Lie group, and to show how it yields the Eells-Wood description [5] of harmonic maps from the sphere into complex projective space. At the same time, I have taken the opportunity to give in Section 1 a simple description of the polynomial loop group of U_n in terms of generators and relations which was suggested by translating the ideas of [10] into the language of Chapter 8 of [9].

Another treatment of some of the material was given by Valli [11], and the theory of harmonic spheres in complex Grassmannians had previously been studied in [12] and [1]. For background and further developments, see [4],[13].

1 A factorization theorem for loops

Let ΩU_n denote the group of based Laurent polynomial loops in the unitary group U_n, i.e. of maps $\gamma: S^1 \to U_n$ such that $\gamma(1) = 1$ which are of the form

$$\gamma(z) = \sum_{k=-N}^{N} A_k z^k$$

for some N, where the A_k are $n \times n$ matrices.

For any vector subspace V of $\mathbf{C}^n$, let λ_V denote the element of ΩU_n given by

$$\lambda_V(z) = \begin{pmatrix} z & 0 \\ 0 & 1 \end{pmatrix},$$

where the matrix is written in terms of the decomposition $\mathbf{C}^n = V \oplus V^\perp$. Clearly,

$$\lambda_V \lambda_W = \lambda_{V \oplus W} \tag{1.1}$$

if V and W are orthogonal.

1.2 Theorem. *The elements λ_V generate the group ΩU_n subject only to the relations* (1.1).

It follows, of course, that the λ_V with $\dim V = 1$ also generate ΩU_n. These form a copy of the complex projective space $P = \mathbf{C}P^{n-1}$ inside ΩU_n.

One can make Theorem 1.2 more precise by introducing the subsemigroup $\check{\Omega}$ of ΩU_n consisting of loops involving only non-negative powers of z. Any $\gamma \in \Omega U_n$ can clearly be written $\gamma = z^{-N}\gamma_0$ for some N, with $\gamma_0 \in \check{\Omega}$. We have

$$\check{\Omega} = \coprod_{k \geq 0} \check{\Omega}_k,$$

where $\check{\Omega}_k$ consists of the loops of winding number k. Theorem 1.2 is a consequence of

1.3 Proposition. (i) $\check{\Omega}_1 = P$.

(ii) *The multiplication* $\underbrace{P \times \cdots \times P}_{k} \to \check{\Omega}_k$ *is surjective, and generically bijective.*

(iii) *Any $\gamma \in \check{\Omega}$ has a factorization*

$$\gamma = \lambda_{V_1}\lambda_{V_2}\cdots\lambda_{V_m}$$

with the property that $V_i^{\perp} \cap V_{i+1} = 0$ for all i. (Thus $\dim V_1 \geq \dim V_2 \geq \cdots \geq \dim V_m$.) This factorization is unique.

Proof. Let H denote the Hilbert space $L^2(S^1; \mathbf{C}^n)$, and let H_+ denote the closed subspace of all elements of the form $\sum_{k \geq 0} \xi_k z^k$, with $\xi_k \in \mathbf{C}^n$. The group ΩU_n acts on H, and it is explained in [9, Chapter 8] that the map $\gamma \mapsto \gamma H_+$ identifies ΩU_n with the set $\mathcal{G}r$ of all subspaces W of H which satisfy the two conditions

(i) $zW \subset W$;

(ii) $z^N H_+ \subset W \subset z^{-N} H_+$ for some N.

The elements λ_V of ΩU_n correspond precisely to the subspaces $W \in \mathcal{G}r$ such that $zH_+ \subset W \subset H_+$. More generally, if $\gamma_1, \gamma_2 \in \Omega U_n$ correspond to $W_1, W_2 \in \mathcal{G}r$, then $\gamma_2 = \gamma_1 \lambda_V$ for some $V \subset \mathbf{C}^n$ if and only if

$$zW_1 \subset W_2 \subset W_1.$$

A factorization $\gamma = \lambda_{V_1} \ldots \lambda_{V_m}$ is therefore the same as a filtration

$$H_+ = W_0 \supset W_1 \supset \cdots \supset W_m = \gamma H_+ \tag{1.4}$$

such that $zW_i \subset W_{i+1}$. Such a filtration obviously exists for any $\gamma \in \check{\Omega}$, and the unique choice leading to $V_i^{\perp} \cap V_{i+1} = 0$ is got by making W_i as small as possible, i.e. taking $W_i = W + z^i H_+$.

To prove the generic bijectivity assertion of 1.3(ii), observe that if $\gamma \in \check{\Omega}_k$ then H_+/W_m is a k-dimensional vector space and also a $\mathbf{C}[z]$-module on which the action of z is nilpotent. It must therefore be annihilated by z^k, but in general is not by z^{k-1}. (Consider, for example, the loop $\gamma = z^k \oplus 1$.) Thus in the generic case $H_+/W_m \cong \mathbf{C}[z]/(z^k)$ as $\mathbf{C}[z]$-modules, and there is no freedom in choosing the filtration (1.4). ∎

We can derive a little more information from the preceding argument. The identification $\Omega U_n \cong \mathcal{G}r$ gives ΩU_n a complex structure, and $\check{\Omega}_k$ is a compact complex analytic algebraic variety with singularities. The multiplication $P \times \cdots \times P \to \check{\Omega}_k$ is, however, not holomorphic, as ΩU_n is not a complex Lie group. The situation can be described as follows.

For each multi-index $\mathbf{a} = (a_1, \ldots, a_m)$, let $F_{\mathbf{a}}$ denote the generalized flag manifold of all sequences

$$H_+ = W_0 \supset W_1 \supset \cdots \supset W_m \tag{1.5}$$

with $zW_i \subset W_{i+1}$ and $\dim(W_{i-1}/W_i) = a_i$. Then $F_{\mathbf{a}}$ is a non-singular algebraic variety, in fact an iterated fibration of Grassmannians. Indeed, as a smooth manifold,

$$F_{\mathbf{a}} \cong \mathrm{Gr}_{a_1}(\mathbf{C}^n) \times \cdots \times \mathrm{Gr}_{a_m}(\mathbf{C}^n),$$

for $W_{i-1} \ominus W_i \hookrightarrow \mathbf{C}^n$ by evaluation at $z = 1$ (see [9, page 126]), but holomorphically the fibrations are non-trivial. The natural holomorphic map $F_{\mathbf{a}} \to \mathcal{G}r$ taking the flag (1.5) to W_m can be identified with the multiplication in the semigroup $\check{\Omega}$.

The most important case is when $\mathbf{a} = 1^k = (1, 1, \ldots, 1)$, and $F_{\mathbf{a}} \cong P \times \cdots \times P$ as a smooth manifold. It is easy to check that $F_{1,1}$ is the bundle of projective spaces on P formed from the vector bundle $T_P \oplus \mathbf{C}$, where T_P is the complex tangent bundle of P, and $\mathbf{C}$ denotes the trivial line bundle. Thus $F_{1,1}$ is a "projective compactification" of the tangent bundle T_P. More generally, F_{1^k} is a compactification of the space of $(k+1)$-jets of holomorphic maps $\mathbf{C} \to P$ (at $0 \in \mathbf{C}$).

If $\check{F}_{\mathbf{a}}$ denotes the image of $F_{\mathbf{a}}$ in ΩU_n, we clearly have

1.6 Proposition. (i) $F_{\mathbf{a}} \to \check{F}_{\mathbf{a}}$ *is a resolution of singularities;*
(ii) $\check{F}_{1^k} = \check{\Omega}_k$;
(iii) $\check{F}_{\mathbf{a}} \subset \check{\Omega}_k$, *where* $k = \sum a_i$; *and more generally*
(iv) $\check{F}_{\mathbf{a}} \subset \check{F}_{\mathbf{b}}$ *if and only if* $\mathbf{b}$ *is a refinement of* $\mathbf{a}$.

To conclude this section, I shall briefly mention one application of the preceding discussion.

The splitting of $\Omega \mathbf{SU_n}$ in stable homotopy theory. Let λ denote λ_V when V is the first coordinate axis of $\mathbf{C}^n$. The space $\check{\Omega}_k$ can be identified with the subspace $X_k^n = \lambda^{-k}\check{\Omega}_k$ of ΩSU_n. The X_k^n for $k = 0, 1, 2, \ldots$ form an expanding sequence whose union is dense in the continuous loop space of SU_n and has the same homotopy type ([9, 8.6.6]). S. Mitchell [8] has used this filtration to analyse the stable homotopy type of ΩSU_n. Notice that $\check{\Omega}_k$, which consists of subspaces W of dimension k in H_+, is naturally a subspace of the Grassmannian $\mathrm{Gr}_k(H_+)$, which has the homotopy type BU_k. In particular, there is a natural $\mathbf{C}^k$-bundle on $\check{\Omega}_k$ whose fibre at W is H_+/W.

1.7 Proposition. *$X_k^n - X_{k-1}^n$ is homeomorphic to the total space of the natural $\mathbf{C}^k$-bundle on X_k^{n-1}.*

In other words X_k^n/X_{k-1}^n is the Thom space of the $\mathbf{C}^k$-bundle on X_k^{n-1}. Mitchell and Richter have proved [2] that the inclusion $X_{k-1}^n \to X_k^n$ is stably split, i.e. that ΩSU_n is stably homotopy equivalent to $\bigvee_{k\geq 0} X_k^n/X_{k-1}^n$. If we let $n \to \infty$ holding k fixed then X_k^n becomes BU_k, so the result is a refinement of Snaith's stable homotopy equivalence

$$BU \sim \bigvee_{k\geq 0} MU_k.$$

Proof of 1.7. Let $A = \mathbf{C}[z]$. Identify X_k^n with the space of A-submodules W of A^n such that $\dim(A^n/W) = k$. Then X_{k-1}^n consists of those W which are contained in $zA \oplus A^{n-1} = zA \oplus M$, say. If $W \in X_k^n - X_{k-1}^n$ then $W + M = A^n$, so $M/(W \cap M) \cong A^n/W$, and $W/(W \cap M)$ is the graph of a homomorphism of A-modules $\varphi: A \to M/(W \cap M)$. The submodule W is completely specified by giving $W \cap M$, which belongs to X_k^{n-1}, and $\varphi \in \mathrm{Hom}_A(A; M/(W \cap M)) \cong M/(W \cap M)$. The pair $(W \cap M, \varphi)$ is a point of the natural $\mathbf{C}^k$-bundle on X_k^{n-1}. ∎

2 Harmonic maps into $\mathbf{U_n}$

We now turn to Uhlenbeck's beautiful description [10] of the harmonic maps from the Riemann sphere to U_n in terms of the complex structure of ΩU_n. It includes and extends the theory of harmonic maps from S^2 to $\mathbf{C}P^{n-1}$ developed in [3] and [5], as we shall see in the next section. The main result is

2.1 Theorem. *Any harmonic map $f: S^2 \to U_n$ has a canonical factorization*

$$S^2 \xrightarrow{\hat{f}} \Omega U_n \xrightarrow{\varepsilon} U_n,$$

where $\hat{f}$ is holomorphic and ε is the evaluation map $\gamma \mapsto \gamma(-1)$. In fact there is a 1–1 correspondence between base-point-preserving harmonic maps $S^2 \to U_n$ and normalized horizontal holomorphic maps $S^2 \to \Omega U_n$. Furthermore for each harmonic f there is a canonical multi-index $\mathbf{a} = (a_1 \leq a_2 \leq \ldots \leq a_m)$, with $m < n$, for which $\hat{f}$ lifts to a horizontal holomorphic map into the generalized flag manifold $F_{\mathbf{a}}$ of Proposition 1.6.

Remarks. (i) The meaning of "horizontal" and "normalized" is explained below.

(ii) The factorization asserted in the theorem has been well known, in some sense, in the Russian literature on integrable systems, but Uhlenbeck's version is more precise, explicit, and useful.

(iii) Identifying the Grassmannian $\mathrm{Gr}(\mathbf{C}^n) = \coprod_{k\geq 0} \mathrm{Gr}_k(\mathbf{C}^n)$ with the elements of order 2 in U_n, the assertion about $F_{\mathbf{a}}$ means that the harmonic map f is canonically a product $f = f_1 f_2 \ldots f_m$, in terms of pointwise multiplication in U_n, where $f_i: S^2 \to \mathrm{Gr}_{a_i}(\mathbf{C}^n)$; but the maps f_i are not holomorphic.

(iv) The theorem implies that the space of based harmonic maps $S^2 \to U_n$ is naturally a complex algebraic variety. This variety breaks into (intersecting) components according to the multi-index $\mathbf{a}$.

(v) Because $H_2(\Omega U_n) = \mathbf{Z}$, a map $\hat{f}: S^2 \to \Omega U_n$ has an integer degree. Concerning this we have the following result (see also [7] and [11]).

2.2 Proposition. *The energy*

$$\mathcal{E}(f) = -\tfrac{1}{2} \int_{S^2} \mathrm{trace}(f^{-1}df \wedge *f^{-1}df)$$

of a harmonic map $f: S^2 \to U_n$ is always an even integer, and is twice the degree of the associated holomorphic map $\hat{f}: S^2 \to \Omega U_n$.

Let us begin with the construction of the holomorphic map $\hat{f}$. If X is a simply connected Riemann surface and $f: X \to U_n$ is a smooth map, we can write

$$f^{-1}df = A_1 dx + A_2 d\bar{x},$$

with x a holomorphic local coordinate on X. The map f is harmonic if and only if

$$\frac{\partial A_1}{\partial \bar{x}} + \frac{\partial A_2}{\partial x} = 0,$$

i.e. if and only if the matrix-valued 1-form $*f^{-1}df$ is closed.

Using the conformal structure of X, it makes sense to rotate a 1-form by an angle θ. When $\alpha = f^{-1}df$ is rotated through θ it becomes

$$\alpha_z = z^{-1}A_1 dx + zA_2 d\bar{x} = \cos\theta.\alpha - \sin\theta. * \alpha,$$

where $z = e^{i\theta}$. We shall consider the circle

$$\begin{aligned}\beta_z &= \tfrac{1}{2}(\alpha - \alpha_z) \\ &= \tfrac{1}{2}(1 - z^{-1})A_1 dx + \tfrac{1}{2}(1 - z)A_2 d\bar{x}.\end{aligned} \tag{2.3}$$

in the space of 1-forms, linking $\beta_1 = 0$ and $\beta_{-1} = \alpha$. A matrix-valued 1-form β is of the form $g^{-1}dg$ for some $g: X \to GL_n(\mathbf{C})$ if and only if it is flat, i.e. $d\beta + \beta \wedge \beta = 0$. We readily verify

2.4 Proposition. *The map f is harmonic if and only if the 1-form β_z is flat for all $z \in \mathbf{C}^\times$.*

Thus if f is harmonic we can find $f_z: X \to GL_n(\mathbf{C})$ for each $z \in \mathbf{C}^\times$ such that $f_z^{-1}df_z = \beta_z$, and f_z is unique up to left-multiplication by a constant element of $GL_n(\mathbf{C})$. We can assume that $f_1 = 1$, $f_{-1} = f$, and that f_z has values in U_n when $|z| = 1$. In other words we have a circle of maps $X \to U_n$ linking f to the constant map 1, or, equivalently, a map $\hat{f}: X \to \Omega_{sm} U_n$, where $\Omega_{sm} U_n$ is the group of smooth based loops $S^1 \to U_n$. The map $\hat{f}$ is unique up to left multiplication by any loop γ such that $\gamma(-1) = 1$. It can be fixed by prescribing it at the base-point of X.

2.5 Lemma. *The map $\hat{f}: X \to \Omega_{sm} U_n$ is holomorphic.*

Proof. The complex structure of $\Omega_{sm} U_n$ comes from its description as $\Omega_{sm} G / \Omega^+_{sm} G$, where $G = GL_n(\mathbf{C})$ and Ω^+ denotes the loops which extend holomorphically to $|z| < 1$. So $\hat{f}$ is holomorphic providing $\hat{f}^{-1}\partial\hat{f}/\partial x$ and $\hat{f}^{-1}\partial\hat{f}/\partial\bar{x}$ belong to $\Omega^-_{sm}\mathfrak{g}$ and $\Omega^+_{sm}\mathfrak{g}$ respectively; and they do so in view of the formula (2.3). ∎

Suppose, on the other hand, that we are given a holomorphic map $\hat{f}: X \to \Omega_{sm} U_n$. If, in terms of a local coordinate x,

$$\hat{f}^{-1}\frac{\partial \hat{f}}{\partial x} = \tfrac{1}{2}(1 - z^{-1})A_1 \tag{2.6}$$

for some matrix-valued function A_1 of x, then $\hat{f}^{-1}d\hat{f}$ is of the form (2.3), and so $\hat{f}$ comes from a harmonic map. The equation (2.6) simply expresses the fact that the tangents to the holomorphic curve $\hat{f}: X \to \Omega_{sm} U_n$ lie in the n^2-dimensional complex subbundle Φ of the tangent bundle of the complex homogeneous space $\Omega_{sm} U_n$ corresponding to the subspace

$$\frac{z^{-1}\Omega^+_{sm}\mathfrak{g}}{\Omega^+_{sm}\mathfrak{g}} \subset \frac{\Omega_{sm}\mathfrak{g}}{\Omega^+_{sm}\mathfrak{g}}.$$

2.7 Definition. *A holomorphic curve in $\Omega_{sm} U_n$ is* horizontal *if its tangents belong to the subbundle Φ.*

At this point we have proved, by a purely local discussion, that there is a 1–1 correspondence between based harmonic maps $f: X \to U_n$ and based horizontal holomorphic maps $\hat{f}: X \to \Omega_{sm} U_n$. We can also give very simply the

Proof of 2.2. The energy $\mathcal{E}(f)$ of $f: S^2 \to U_n$ is defined by

$$\begin{aligned}\mathcal{E}(f) &= -\tfrac{1}{2}\int_{S^2} \text{trace}(f^{-1}df \wedge *f^{-1}df) \\ &= -i\int_{S^2} \text{trace}(A_1 A_2)dx \wedge d\bar{x}.\end{aligned}$$

On the other hand the generator of $H^2(\Omega U_n)$ is represented by the left-invariant 2-form ω given on the Lie algebra $\Omega\mathfrak{u}_n$ by

$$\omega(\xi,\eta) = \frac{1}{2\pi}\int_{S^1} \text{trace}(\xi d\eta).$$

When ω is pulled back to S^2 by $\hat{f}$ it becomes

$$\frac{1}{2\pi}\int_{S^1} \text{trace}(\beta_z \frac{d}{dz}\beta_z)dz$$
$$= \frac{1}{8\pi}\text{trace}(A_1A_2)dx \wedge d\bar{x} \int_{S^1} \left\{(1-z^{-1})d(1-z) - (1-z)d(1-z^{-1})\right\}$$
$$= \frac{i}{2}\text{trace}(A_1A_2)dx \wedge d\bar{x}.$$

Thus $\mathcal{E}(f)$ is twice the degree of $\hat{f}$.

To complete the proof of Theorem 2.1, we must show that when $X = S^2$ the image $\hat{f}(S^2)$ is contained in the Laurent polynomial loop group, and, better still, that $\hat{f}$ can be translated by a canonical element of ΩU_n so that $\hat{f}(S^2)$ is contained in one of the generalized flag manifolds $\check{F}_{\mathbf{a}}$ of length less than n.

We shall use the Grassmannian description of $\Omega_{\text{sm}}U_n$ The holomorphic map $\hat{f}\colon S^2 \to \Omega_{\text{sm}}U_n$ is identified with a map $S^2 \to \mathcal{G}r_{\text{sm}}$ which will be written $x \mapsto W_x$. The horizontality condition translates into*

$$\frac{\partial W_x}{\partial x} \subset z^{-1}W_x. \tag{2.8}$$

The result we need is

2.9 Proposition. *Let X be a compact complex manifold. If $x \mapsto W_x$ is a holomorphic map $X \to \mathcal{G}r_{\text{sm}}$ then there is an element $U \in \mathcal{G}r_{\text{sm}}$ such that $W_x \subset U$ for all x.*

Postponing the proof for a moment, let us see how 2.9 implies the remainder of Theorem 2.1. First we have

2.10 Corollary. *The holomorphic map $\hat{f}\colon S^2 \to \Omega_{\text{sm}}U_n$ can be normalized canonically so that $\bigcup W_x$, for $x \in S^2$, spans H_+.*

This is the *Uhlenbeck normalization*, which we shall presuppose from now on. It replaces the condition that $\hat{f}$ is base-point preserving.

2.11 Lemma. *For some integer m we have $z^m H_+ \subset W_x$ for all $x \in S^2$.*

The least such m is called the *uniton number* of f. The lemma implies that $\hat{f}(S^2)$ consists of Laurent polynomial loops.

---

* If W_x and V_x are two varying subspaces of a vector space H, parametrized by $x \in X$, then the notation $\partial W_x/\partial x \subset V_x$ means that whenever $w\colon X \to H$ is such that $w(x) \in W_x$ for all x, one has $\partial w/\partial x \in V_x$.

Proof of 2.11. For each x the vector space $V_x = H_+/W_x$ is finite dimensional. Multiplication by z defines an endomorphism of V_x. Let its characteristic polynomial be $p_x \in \mathbb{C}[z]$. Because p_x depends holomorphically on x it is actually independent of x. Thus we have $p \in \mathbb{C}[z]$ such that $pH_+ \subset W_x$ for all x. It follows that the loop $\hat{f}(x)$ corresponding to W_x is a rational function of z with poles precisely at the zeros of p. But $\hat{f}(x)$ extends by definition to a holomorphic map $\mathbb{C}^\times \to GL_n(\mathbb{C})$. So $p = z^m$ for some m.

2.12 Proposition. *The uniton number m is less than n.*

Proof. We have

$$H_+ = W_{x,0} \supset W_{x,1} \supset \cdots \supset W_{x,m} = W_x, \tag{2.13}$$

where $W_{x,m-i} = z^{-i}W_x \cap H_+$. Each of these spaces depends holomorphically on x except perhaps at a finite set of points (where the dimensions jump). Now define $W'_{x,j} = W_{x,j} + zH_+$, so that

$$H_+ \supset W'_{x,1} \supset W'_{x,2} \supset \cdots \supset W'_{x,m} \supset zH_+. \tag{2.14}$$

Because $\partial W_x/\partial x \subset z^{-1}W_x$ we have $\partial W_{x,j}/\partial x \subset W_{x,j-1}$ and $\partial W'_{x,j}/\partial x \subset W'_{x,j-1}$. But no $W'_{x,j}$ can be constant (unless it is H_+), for $W'_{x,j} \supset W_x$, and the W_x span H_+. It follows that all the inclusions in (2.14) are strict, and hence that $m < n = \dim(H_+/zH_+)$.

We have now proved everything that was wanted, for the map which takes x to the canonical flag (2.13) is a holomorphic map $S^2 \to F_{\mathbf{a}}$, where $a_i = \dim(W_{x,i}/W_{x,i+1})$. It is horizontal in the sense that

$$\frac{\partial W_{x,i}}{\partial x} \subset W_{x,i-1}$$

for each i. (Although the map was at first undefined at a finite set of points it can be extended over all of S^2 because it is algebraic and $F_{\mathbf{a}}$ is complete. It is algebraic because it is obtained algebraically from the holomorphic map $\hat{f}$ from S^2 into the finite dimensional Grassmannian $\mathrm{Gr}(H_+/z^m H_+)$.)

It remains to give the omitted proof of 2.9.

2.15 Lemma. *If X is a compact complex manifold, E is a locally convex topological vector space, and $\varphi: X \to P(E)$ is a holomorphic map, then $\varphi(X) \subset P(F)$, where F is a finite dimensional subspace of E.*

Proof. Let $L_X = \varphi^* L$, where L is the canonical line bundle on $P(E)$, whose space of holomorphic sections is E^*. The map φ is completely described by the adjoint map $\varphi^*: E^* \to \Gamma(L_X)$, and $\varphi(X)$ is contained in $P(F)$, where F is the annihilator of the kernel of φ^*. But F is finite dimensional because $\Gamma(L_X)$ is finite dimensional.

2.16 Corollary. *In the situation of 2.15, if $\psi: X \to \mathrm{Gr}_k(E)$ is a holomorphic map into the Grassmannian of k-dimensional subspaces of E, then $\psi(X) \subset \mathrm{Gr}_k(F)$ for some finite-dimensional $F \subset E$.*

Proof. Apply 2.15 to $\varphi: Y \to P(E)$, where Y is the compact manifold of pairs (x, λ) with $x \in X$ and λ a line in $\psi(x)$, and $\varphi(x, \lambda) = \lambda$. ∎

Proof of 2.9. Let $H_q = z^q H_+ \subset H$. For any q, the subset $\mathcal{G}r^{(q)}$ of $\mathcal{G}r_{\mathrm{sm}}$ consisting of W such that $W + H_q^\perp = H$ is open. Because X is compact we can find q so that $W_x \in \mathcal{G}r^{(q)}$ for all $x \in X$. Then $x \mapsto W_x \cap H_q^\perp$ is a holomorphic map into $\mathrm{Gr}_k(H_q^\perp)$ for some finite k, so by 2.16, there is a finite dimensional subspace F of $H_q^\perp$ such that $W_x \cap H_q^\perp \subset F$ for all x. Let $K = H_q^\perp \ominus F$. Then $W_x + F \in \mathcal{G}r$ for all x, and is transversal to K, for $(W_x + F) + K = W_x + H_q^\perp = H$, while

$$\begin{aligned}(W_x + F) + K &= (W_x + F) \cap H_q^\perp \cap K \\ &\subset F \cap K = 0.\end{aligned}$$

So $W_x + F$ is the graph of a compact operator from $K^\perp$ to K. This operator depends holomorphically on x, so is independent of x. Then $U = W_x + F$ is independent of x and contains W_x.

3 Harmonic maps into $\mathbf{C}P^{n-1}$

The Eells-Wood theorem [5] describes the harmonic maps $f: S^2 \to P = \mathbf{C}P^{n-1}$ in the following way. Let $\varphi: S^2 \to P$ be a holomorphic curve. For each k with $1 \le k \le n-2$, there is an induced holomorphic map $\varphi_k: S^2 \to \mathrm{Gr}_k(\mathbf{C}^n)$ characterized by $\varphi_1 = \varphi$ and $\varphi_k'(x) \subset \varphi_{k+1}(x)$. (In other words, $\varphi_k(x)$ is the osculating $(k-1)$-plane to the curve φ at $\varphi(x)$.) Define $\psi_k: S^2 \to P$ by $\psi_k(x) = \varphi_k(x) \ominus \varphi_{k-1}(x)$. Then ψ_k is harmonic, and all harmonic maps are so obtained. We shall now show how this theorem follows from Uhlenbeck's Theorem 2.1.

The set of elements of order 2 in U_n can be identified with the Grassmannian $\mathrm{Gr}(\mathbf{C}^n) = \coprod_{k \ge 0} \mathrm{Gr}_k(\mathbf{C}^n)$. Because it is the fixed points of the involution $g \mapsto g^{-1}$, the Grassmannian is totally geodesic in U_n, and so a harmonic map into $\mathrm{Gr}(\mathbf{C}^n)$ is the same as a harmonic map into U_n whose image is contained in $\mathrm{Gr}(\mathbf{C}^n)$. In the language of Section 2, the Eells-Wood theorem amounts to

3.1 Theorem. *If $f: S^2 \to U_n$ is a harmonic map whose image is contained in $P = \mathrm{Gr}_1(\mathbf{C}^n)$, then it has uniton number ≤ 2.*

We shall explain below how 3.1 leads to the formulation given above, but first we shall derive 3.1 from 2.1.

Suppose that the harmonic map f corresponds to the holomorphic map $x \mapsto W_x$ into $\mathcal{G}r$, normalized in Uhlenbeck's sense. There is a holomorphic involution T on

$\mathcal{G}r$ induced by the map $z \mapsto -z$ on S^1. It corresponds to $\gamma \mapsto \tilde{\gamma}$ on ΩU_n, where $\tilde{\gamma}(z) = \gamma(-z)\gamma(-1)^{-1}$. The evaluation $\varepsilon: \mathcal{G}r \to U_n$ is equivariant in the sense that $\varepsilon(T.W) = \varepsilon(W)^{-1}$.

Because $x \mapsto W_x$ is canonically associated to f, it follows that W_x is invariant under T for all x. So

$$W_x = W_x^{\text{even}} \oplus W_x^{\text{odd}},$$

where "even" and "odd" mean that T acts as $+1$ and -1 respectively. If $W_x \ominus zW_x$ is identified with $\mathbf{C}^n$ by evaluating at $z = 1$ then the element of order 2 in U_n corresponding to W_x is given by the decomposition

$$W_x \ominus zW_x = (W_x \ominus zW_x)^{\text{even}} \oplus (W_x \ominus zW_x)^{\text{odd}}.$$

Because $f(S^2) \subset P$ we conclude that $\dim(W_x \oplus zW_x)^{\text{odd}} = 1$.

If f has uniton number m then $H_0 \supset W_x \supset H_m$, where $H_i = z^i H_+$. The filtration of W_x by the $W_x \cap H_i$ induces a splitting

$$W_x \ominus zW_x = A_0 \oplus A_1 \oplus \cdots \oplus A_m,$$

where $A_i = (W_x \cap H_i) \ominus ((zW_x \cap H_i) + (W_x \cap H_{i+1}))$. This splitting is stable under T, and T acts on A_i as $(-1)^i$. We therefore have

$$\sum_{i \text{ odd}} \dim A_i = 1.$$

To complete the proof that $m \leq 2$, it suffices to show that $\dim A_i > 0$ for $i \leq m$. But

$$\begin{aligned} A_i &\cong (z^{-i}W_x \cap H_0)/((z^{-i+1}W_x \cap H_0) + (z^{-i}W_x \cap H_1)) \\ &\cong ((z^{-i}W_x \cap H_0) + H_1)/((z^{-i+1}W_x \cap H_0) + H_1) \\ &= W'_{x,m-i}/W'_{x,m-i+1}, \end{aligned}$$

in the notation of (2.14). But we proved that this space was non-zero. ∎

In the language of Theorem 2.1, we have now proved that $\hat{f}: S^2 \to \Omega U_n$ lifts into the flag manifold $F_{r,r+1}$ for some r. Because it is T-invariant, the flag

$$H_0 \supset W_{x,1} \supset W_x \supset H_2$$

of (2.13) must be of the form

$$\begin{aligned} W_{x,1} &= B_x \oplus H_1 \\ W_x &= A_x \oplus zB_x \oplus H_2, \end{aligned}$$

where A_x and B_x are subspaces of $\mathbf{C}^n$ of dimensions $n - r$ and $n - r + 1$ respectively such that $A_x \subset B_x$. Both A_x and B_x depend holomorphically on x, and the map $f: S^2 \to P$ is given by $f(x) = B_x \ominus A_x$. To obtain the original formulation of the Eells-Wood theorem, it is therefore enough to prove the following result (see also [6, prop. 2.3]).

3.2 Proposition. *Let Σ be a closed Riemann surface. If $x \mapsto A_x$ is a holomorphic map $\Sigma \to \mathrm{Gr}_k(\mathbf{C}^n)$ such that $\partial A_x/\partial x$ has dimension $k+1$, then $A_x = \varphi_k(x)$, the kth osculating space of a holomorphic curve $\varphi: \Sigma \to P$.*

This, in turn, will be deduced from

3.3 Lemma. *In the situation of 3.2, we have $\dim(\partial^r A_x/\partial x^r) \leq k+r$, with equality except at a finite set of points $x \in \Sigma$ unless all A_x are contained in a subspace of $\mathbf{C}^n$ of dimension less than $k+r$.*

Proof of 3.3. In an open set U of Σ, choose holomorphic functions $\xi_1, \ldots, \xi_k, \eta: U \longrightarrow \mathbf{C}^n$ such that the set $\{\xi_1(x), \ldots, \xi_k(x)\}$ spans A_x and the set $\{\xi_1(x), \ldots, \xi_k(x), \eta(x)\}$ spans $A'_x = \partial A_x/\partial x$. Then A''_x is spanned by the set $\{\xi_1(x), \ldots, \xi_k(x), \eta(x), \eta'(x)\}$, and so on. If $(A_x^{(i)})' \subseteq A_x^{(i)}$ for more than a finite set of points x, then $A_x^{(i)}$ must be independent of x, and is a $(k+i)$-dimensional subspace of $\mathbf{C}^n$ which contains all A_x. ∎

Proof of Proposition 3.2. In the dual space of $\mathbf{C}^n$, let A_x^{ann} denote the annihilator of A_x, and let $\{\alpha_1(x), \ldots, \alpha_r(x), \beta(x)\}$ be a basis for it such that $\{\alpha_1(x), \ldots, \alpha_r(x)\}$ is a basis for $(A'_x)^{\mathrm{ann}}$ (thus $r = n-k-1$). Then $\alpha'_i(x) \in A_x^{\mathrm{ann}}$, so $(A_x^{\mathrm{ann}})'$ is spanned by $\{\alpha_1(x), \ldots, \alpha_r(x), \beta(x), \beta'(x)\}$, and has dimension $n-k+1$. Applying Lemma 3.3 to A_x^{ann} we can assume that $\dim(A_x^{\mathrm{ann}})^{(k-1)} = n-1$. Let L_x be the annihilator of $(A_x^{\mathrm{ann}})^{(k-1)}$. Then $\dim L_x = 1$, and $x \mapsto L_x$ is a holomorphic curve $\varphi: \Sigma \to P$. But if L_x annihilates $(A_x^{\mathrm{ann}})^{(k-1)}$, then $L_x^{(k-1)}$ annihilates A_x^{ann}. So $L_x^{(k-1)} = A_x$, as we want.

References

[1] F.E. Burstall and S.M. Salamon: Tournaments, flags, and harmonic maps. *Math. Ann.* 277 (1987), 249–265.

[2] M.C. Crabb and S.A. Mitchell: The loops on $U(n)/O(n)$ and $U(2n)/Sp(n)$. *Math. Proc. Cambridge Philos. Soc.* 104 (1988), 95–103.

[3] A.M. Din and W.J. Zakrzewski: General classical solutions of the CP^{n-1} model. *Nuclear Phys.* B 174 (1980), 397–406.

[4] J. Eells and L. Lemaire: Another report on harmonic maps. *Bull. London Math. Soc.* 20 (1988), 385–524.

[5] J. Eells and J.C. Wood: Harmonic maps from surfaces to complex projective spaces. *Adv. in Math.* 49 (1983), 217–263.

[6] S. Erdem and J.C. Wood: On the construction of harmonic maps into a Grassmannian. *J. London Math. Soc.* 28 (1983), 161–174.

[7] D.S. Freed: The Geometry of Loop Groups. Thesis, M.I.T. 1985.

[8] S.A. Mitchell: The filtration of the loops on $SU(n)$ by Schubert varieties. *Math. Z.* 193 (1986), 347–362.

[9] A. Pressley and G. Segal: Loop Groops. Oxford Mathematical Monographs, Clarendon Press, 1986.

[10] K. Uhlenbeck: Harmonic maps into Lie groups (Classical solutions of the chiral model). Preprint, University of Chicago, 1985.

[11] G. Valli: On the energy spectrum of harmonic 2-spheres in unitary groups. *Topology* 27 (1988), 129–136.

[12] J.G. Wolfson: Harmonic sequences and harmonic maps of surfaces into complex Grassmann manifolds. *J. Differential Geom.* 27 (1988), 161–178.

[13] J.C. Wood: Explicit construction and parametrization of harmonic two-spheres in the unitary group. *Proc. London Math. Soc.* 58 (1989), 608–624.

An almost groupoid structure for the space of (open) strings and implications for string field theory

Jim Stasheff*

Introduction

One version of string theory considers a string as a geometric object — something like an oriented arc in a Riemannian manifold $(\mathcal{M}, g)$. If smooth, a curve C has a length given by $L(C) = \int_0^r d\sigma \sqrt{g_{\mu\nu} X^{\mu\prime}(\sigma) X^{\nu\prime}(\sigma)}$ which is independent of the choice of parameterization $X\colon [0, r] \to \mathcal{M}$ of C. This is prototypical of one of the subtleties of string theory: using parameterizations X to obtain parameterization-invariant results.

String interactions are handled by "joining" two strings to form a third: this is often pictured in one of three ways:

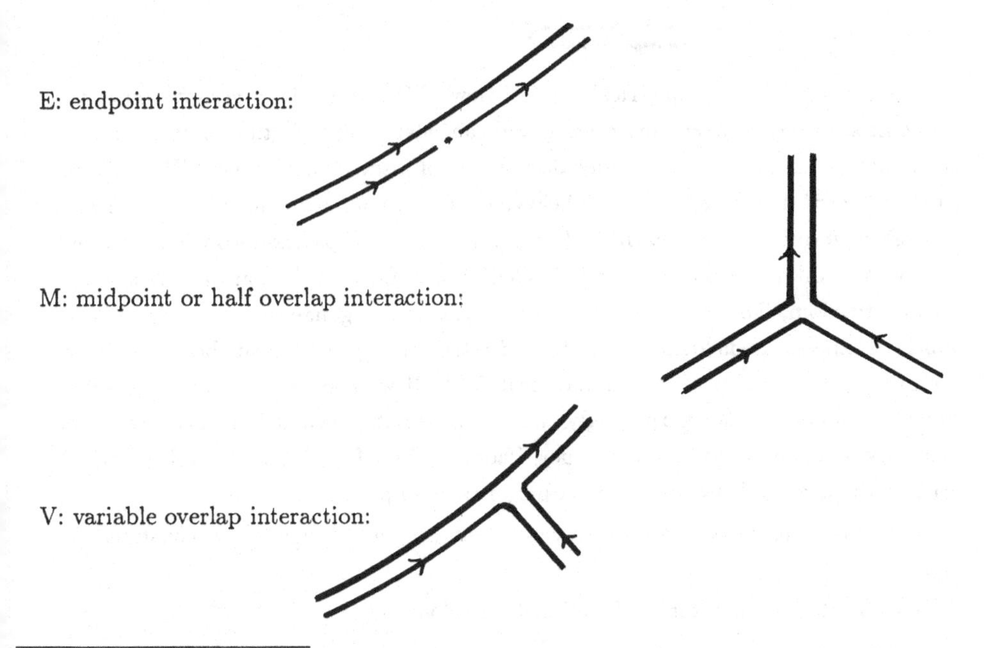

*Research supported in part by National Science Foundation Grant DMS-8506637

The endpoint case E is familiar in mathematics as far back as the study of the fundamental group; the midpoint case M was considered by [Lashof]; the variable case V seems to have occurred first in physics in the work of [Kaku]. Warning: Physicists treat the three strings more symmetrically by reversing the orientation of the "composite", e.g.,

These operations on $\mathcal{S}$, the space of strings, are all meant to make $\mathcal{S}$ into a group or something like it. Indeed, the appropriate mathematical concept is that of groupoid. The major point is that only pairs (C_1, C_2) such that the end of C_1 is the beginning of C_2 can be joined. We prefer the terminology: source for beginning and target for end. The first two of the above operations are associative: three strings end to end determine a unique total string. Notice in the midpoint case, the middle string disappears completely. In contrast to group multiplication, there are many "units" for string joining: the point strings of length 0 at any point of $\mathcal{M}$ in cases E and V. Inverses are more subtle. The obvious candidate for the inverse of C is the same subset of $\mathcal{M}$ but with the reverse orientation: the picture 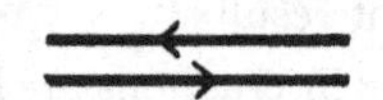

shows this to be correct for case V but also for the midpoint case since there the units are the paths 

From the physicist's symmetrical point of view, this picture also gives inverses in the endpoint gauge by reversing orientation and cancellation of a portion of one string by all of another. This leads to a hidden destruction of associativity in what [HIKKO] refer to as "horn" diagrams; see also §3 below. In any case, for string field theory, we wish to work in terms of *parameterized* strings but produce a parameterization-invariant theory. As [Kaku] has pointed out very clearly, we can treat the parameterization as a gauge freedom. Then we have the choice of choosing a gauge (parameterization) in which to express the fields or trying to find a true string gauge field theory — where the gauge parameter is an additional local field. If we choose different gauges (parameterizations), we not surprisingly can obtain results, even BRST theories, which look very different. In the present paper, I address three foundational questions which arise at the level of $\mathcal{PM}$, the space of parameterized paths $X\colon [0,r] \to \mathcal{M}$:

1) the breaking of associativity in case V, and in the "gauge" $r =$ constant in case E;

2) a convolution operator for functionals on $\mathcal{PM}$;

3) tensor calculus over $\mathcal{PM}$.

What follows is based on two versions of string field theory, as I have encountered it in the physics literature, particularly in the work of the Kyoto group: [HIKKO]: Hata, Itoh, Kugo, Kunitomo, and Ogawa, on the one hand, and [Kaku] on the other. I have attempted to find a common mathematical foundation for a combination of their approaches.

1 The Moore space of paths and various joining operations

For many years, topologists worked with the space of (continuous) maps $\mathrm{Map}(I, \mathcal{M})$ where $I = [0,1]$. (Physicists prefer $[0,\pi]$.) Here endpoint joining E requires a reparameterization:

$$X+Y: \begin{array}{lll} t \mapsto X(2t) & \text{for} & 0 \le 2t \le 1 \\ t \mapsto Y(2t-1) & \text{for} & 1 \le 2t \le 2. \end{array}$$

This operation failed to have units, inverses or associativity, though all were present "up to homotopy". In 1956, [Moore] constructed his space of loops $X: [0,r] \to \mathcal{M}$ such that $X(0) = X(r) = m_0$ so as to have strict associativity and unit (although not inverses) — what is known as a topological monoid. Removing the basepoint restriction, we can consider the space of parameterized paths $X: [0,r] \to \mathcal{M}$, topologized as a subspace of $\mathbf{R}^+ \times \mathrm{Map}(I, \mathcal{M})$ where X is imbedded as $(r, \underline{X})$ with $\underline{X}(t) = X(rt)$. Fixing r can be regarded as a (partial) choice of gauge. Now we define the *endpoint* joining operation $X \vee Y$ for X,Y in $\mathcal{PM}$ such that X(r) = Y(0) with $X \vee Y : [0, r+s] \to \mathcal{M}$ by

$$\begin{array}{lll} t \mapsto X(t) & \text{for} & 0 \le t \le r \\ t \mapsto Y(t-r) & \text{for} & r \le t \le r+s. \end{array}$$

This operation is associative where defined and has units $m: [0,0] \to \mathcal{M}$, but has inverses only up to homotopy. Moore shows the operation is continuous.

[Lashof]'s half cancellation operation generalizes to "midpoint joining" of paths X and Y in $\mathrm{Map}(I, \mathcal{M})$ such that $X(t) = Y(1-t)$ for $1/2 \le t \le 1$:

$$X \bowtie Y: \begin{array}{lll} t \mapsto X(t) & \text{for} & 0 \le t \le 1/2 \\ t \mapsto Y(t) & \text{for} & 1/2 \le t \le 1. \end{array}$$

This has strict units (not the point paths), inverses (given by reversing orientation), and associativity, but is hopelessly parameterization dependent — the midpoint is defined in terms of the parameterization. It is also troublesome that the triple join $X \bowtie Y \bowtie Z$ does not depend on Y!

The variable overlap joining V leads to a different operation $X*Y$ defined as follows. Let $u = \max\{r : Y(t) = X(r-t) \text{ for } 0 \le t \le r\}$. Then $X * Y: [0, r+s-2u] \to \mathcal{M}$ is given by

$$
\begin{aligned}
t &\mapsto X(t) && \text{for} \quad 0 \le t \le r-u \\
t &\mapsto Y(t-r+2u) && \text{for} \quad r-u \le t \le r+s-2u.
\end{aligned}
$$

Again $m: [0,0] \to \mathcal{M}$ is a unit, and the reversal $\overline{X}(t) = X(r-t)$ provides an inverse, but $*$ occasionally fails to be associative; for example in the configuration

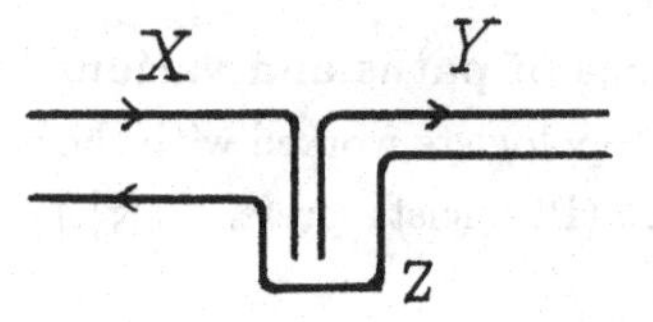

we have

$(X * Y) * Z =$ and $X * (Y * Z) =$.

On the other hand, $\vee$ and $*$ are clearly homotopic operations — just use the formulas for $X * Y$ with v in place of u where $0 \le \mathrm{v} \le u$ and hence $*$ is homotopy associative. From our point of view, it is this homotopy which gives rise to the "4-string vertex" in physics.

2 Convolution algebras

String field theory includes consideration of functionals $\Phi, \Psi: \mathcal{PM} \to \mathbf{R}$ (or $\mathbf{C}$). If $\mathcal{PM}$ were a compact group G, we could define the convolution $\Phi * \Psi$ by

$$
(\Phi * \Psi)(g) = \int_G \Phi(gh)\Psi(h^{-1})d\mu
$$

where $d\mu$ is an invariant measure on G with $\int_G d\mu = 1$. Instead we define

$$
(\Phi * \Psi)(Z) = \int_{X*Y=Z} \Phi(X)\Psi(Y)d\mu
$$

where $d\mu$ is an appropriate measure on $\mathcal{PM}$, e.g., Wiener measure. Notice that since we are decomposing Z in all possible ways under $*$, the integration is over all paths W that have their source at a point of Z.

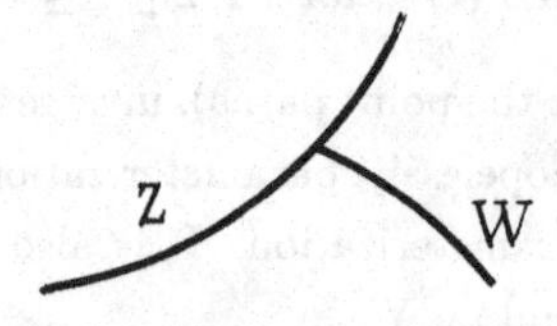

This also shows that if Φ, Ψ are reparameterization invariant, so is $\Phi * \Psi$. (Un)fortunately, $*$ is not associative for paths and this failure carries over to $*$ for functionals.

This gives rise to an operation $\Phi \circ \Psi \circ \Lambda$ for three functionals, which plays the role of an associating homotopy.

The operation $\vee$ is associative but lacks inverses. Physicists force inverses in the endpoint gauge by interpreting

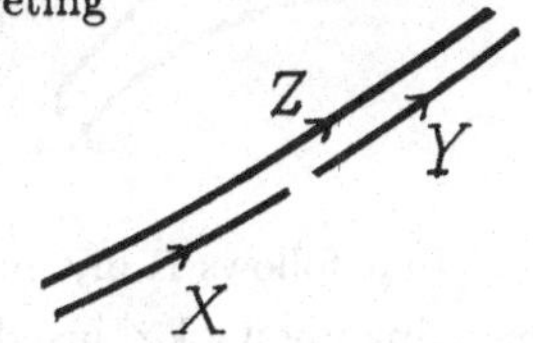

as $Z \vee \overline{Y} = X$, but this again leads to a failure of associativity; indeed, this leads to the variable overlap operation $*$ and hence to the operation $\Phi \circ \Psi \circ \Lambda$. Without thinking of $\Phi \circ \Psi \circ \Lambda$ as an associating homotopy and certainly without knowledge of A_4-structures [Stasheff], HIKKO considers the appropriate combination

$$\Phi * (\Psi \circ \Lambda \circ \Sigma) - (\Phi \circ \Psi \circ \Lambda) * \Sigma - (\Phi * \Psi) \circ \Lambda \circ \Sigma \\ + \Phi \circ (\Psi * \Lambda) \circ \Sigma - \Phi \circ \Psi \circ (\Lambda * \Sigma)$$

and assert that it vanishes — at least if the dimension of M is 26. The general theory of A_∞-structures as applied to $\mathcal{PM}$ would predict a hierarchy of such "n-string interactions".

3 Tensor calculus on string space

String field theory also looks at sections of bundles E over $\mathcal{S}$. One of Kaku's key insights is to treat $\mathcal{PM} \to \mathcal{S}$ as a principal bundle and mimic gauge field theory, e.g., Yang-Mills. Then E in turn can be studied via an associated bundle $\mathcal{E}$ with appropriate "gauge degrees of freedom" over $\mathcal{PM}$. The group $\mathcal{G}$ of the bundle $\mathcal{PM} \to \mathcal{S}$ is the group of reparameterizations

$$\mathcal{G} = \mathrm{Diff}^+ I \propto \mathbf{R}^+$$

where $\propto$ denotes the semidirect product, $\mathbf{R}^+$ acting by changing scale: for $X\colon [0,r] \to \mathcal{M}$, $\phi \in \mathrm{Diff}^+ I$, $s \in R^+$, we have $(\phi, s)X\colon [0, sr] \to \mathcal{M}$ by $t \mapsto X(r\phi(t/sr))$. This in turn allows Kaku to treat the "tangent bundle to $\mathcal{S}$" in terms of a bundle over $\mathcal{PM}$, but *not* the obvious tangent bundle $T(\mathcal{PM}) = \mathcal{P}T\mathcal{M}$ in which a tangent vector to a path is a path of tangent vectors in $\mathcal{M}$.

The point is that since a string is an equivalence class of parameterized paths, a tangent vector to the space of strings is an infinitesimal family of such equivalence classes or an equivalence class of infinitesimal families of paths. The appropriate picture is an infinitesimal strip

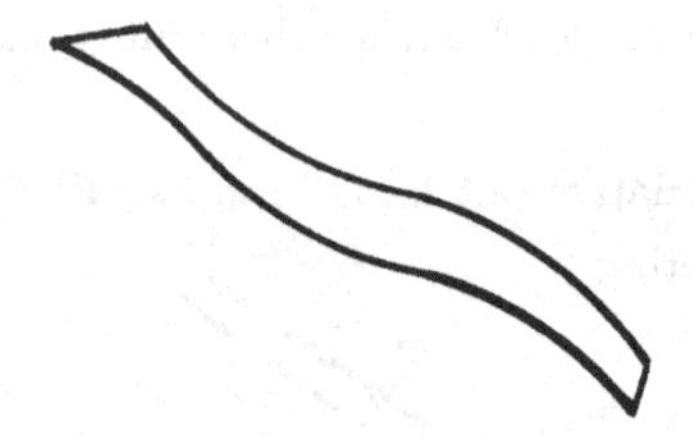

in M, *not* a vector field along the path. What follows is my interpretation of Kaku's formulas for handling this situation, depending upon a key insight due to Paolo Cotta-Ramusino, based on the Polyakov "action" in string field theory.

Consider the space $\mathcal{UPM}$ of bundle maps $T[0,r] \to T\mathcal{M}$ which projects naturally onto $\mathcal{PM}$. For smooth paths in $\mathcal{PM}$, there is a natural lift into $\mathcal{UPM}$, but being concerned with (infinitesimal) variations in the path, both transverse and longitudinal, we need the more general bundle maps. Similarly, the usual tangent space $T\mathcal{PM} = \mathcal{P}T\mathcal{M}$ maps into $\mathcal{UPM}$, but this amounts to choosing a section or trivialization of T[0,r] and we need the extra freedom of $\mathcal{UPM}$ to see the infinitesimal effect of reparameterization. Indeed the action of $\mathrm{Diff}^+ I$ induces an action on T[0,1] and this extends to an action of $\mathcal{G}$ on the family T[0,r] and thence to an action on $\mathcal{UPM}$. The tangent bundle $T\mathcal{S}$ can be defined as the quotient $\mathcal{UPM}/\mathcal{G}$. (The tangent space to $\mathcal{PM}$ at a path X can be identified with the space of sections $X^*T\mathcal{M}$ whereas the tangent space to $\mathcal{S}$ at the equivalence class C = [X] can be identified with the space of bundle maps $T[0,r] \to X^*T\mathcal{M}$ modulo the action of $\mathcal{G}$.) Aside: We can choose to identify T[0,r] with $[0,r] \times \mathbf{R}$ and thus arrive at an alternative interpretation of "world sheet coordinates" $(\sigma,\tau), \sigma \in [0,r], \tau \in \mathbf{R}$.

Having a tangent bundle, we can proceed to tensor calculus in terms of "tensor fields" Φ over $\mathcal{PM}$, that is to say, sections Φ over $\mathcal{PM}$ of

$$\mathcal{UPM} \otimes \cdots \otimes \mathcal{UPM} \otimes U^*\mathcal{PM} \otimes \cdots \otimes U^*\mathcal{PM}.$$

To "raise and lower indices", we need to identify $\mathcal{UPM}$ with $U^*\mathcal{PM}$ fibre-wise and in a way that is reparameterization-invariant. Kaku's solution is to use a "vierbein" e which I regard as a "metric" g^{ij} parameterized by T[0,r]. As an ordinary metric is a section of the symmetric tensor product $T^*\mathcal{M} \odot T^*\mathcal{M}$, so we use bundle maps $e\colon T[0,r] \to U^*\mathcal{PM} \odot U^*\mathcal{PM}$ which vary appropriately under the action of $\mathcal{G}$. Thus Cotta-Ramusino and I interpret Kaku's formulas for an inner product on the space of sections of $\mathcal{UPM}$. The standard formalism of Lagrangians and/or Hamiltonian mechanics is now available.

Kaku also arrives at "4-point vertices" which correspond to homotopy associativity of $*$ and, like the Kyoto group, asserts that higher order terms are unnecessary — at least in $D = 26$.

But for those who tolerate (even enjoy) higher order associativity, maybe we need not live in 26 dimensions after all!

Postscript. After my Cortona talk and further study of HIKKO and Kaku for closed strings, I returned to [Witten]'s "Non-commutative geometry and string field theory". His $*$ operator can be regarded as my mid-point operator $\bowtie$ with an additional factor depending on $Z(\frac{\pi}{2})$. Apparently this factor compensates for the manifest reparameterization NON-invariance of the operation. A 4-point function is obtained by integration over a parameter; this translates as follows: The "s-channel" for four paths can be read as $(Z_1 + Z_2) + Z_3 = \overline{Z}_4$ while the "t-channel" reads $Z_1 + (Z_2 + Z_3) = \overline{Z}_4$. The parameter provides an associating homotopy which the integration converts to a chain homotopy at the form level.

On further examination, HIKKO's proof of the A_4 identity seems to be a consequence of the fact that the associating homotopy involved can be expressed in terms of the homotopy showing that $\overline{X}$ is homotopy inverse to X.

References

HIKKO H. Hata, K. Itoh, T. Kugo, H. Kunitomo and K. Ogawa: Covariant string field theory. *Physical Review* D 34 (1986), 2360-2429.

HIKKO H. Hata, K. Itoh, T. Kugo, H. Kunitomo and K. Ogawa: Covariant string field theory. II. *Physical Review* D 35 (1987), 1318-1355.

HIKKO T. Kugo: String Field Theory. *The SuperWorld II, Lectures delivered at 25th Course of the International School of Subnuclear Physics, Erice*, August, 1987.

M. Kaku: Why are there two BRST string field theories? *Phys. Lett.* B 200 (1988) 22-30.

M. Kaku: Deriving the four-string interaction from geometric string field theory. *Preprint, CCNY-HEP-88/5.*

M. Kaku: Geometric derivation of string field theory from first principles: Closed strings and modular invariance. *Preprint, CCNY-HEP-88/6.*

M.Kaku and J. Lykken: Modular invariant closed string field theory. *Preprint, CCNY-HEP-88/7.*

R. Lashof: Classification of fibre bundles by the loop space of the base. *Ann. of Math.* 64 (1956) 436-446.

J.C. Moore: The double suspension and p-primary components of the homotopy groups of spheres. *Bol. Soc. Mat. Mexicana* 1 (1956) 28-37.

J. Stasheff: Homotopy associativity of H-spaces. I. *Trans. Amer. Math. Soc.* 108 (1963), 275-292.

J. Stasheff: Homotopy associativity of H-spaces. II. *Trans. Amer Math. Soc.* 108 (1963), 293-312.

J. Stasheff: H-Spaces from a homotopy point of view. Lecture Notes in Math. 161, Springer, Heidelberg-New York, 1970.

E. Witten: Non-commutative geometry and string field theory. *Nuclear Phys.* B268 (1986), 253-294.

E. Witten: Interacting field theory of open strings. *Nuclear Phys.* B276 (1986) 291-324.

Steenrod squares of polynomials

R.M.W. Wood

1 The Peterson Conjecture

About two years ago Frank Peterson made a conjecture about the degrees d of generators for the polynomial algebra

$$S = F_2[x_1, \ldots, x_n]$$

in n variables, viewed as a module over the Steenrod algebra A_2, where S is identified with the cohomology of the product of n copies of RP^∞ with coefficients in F_2, the field of two elements. To explain the Peterson Conjecture we introduce some terminology by calling d a *Peterson n-number* if

$$d = \sum_{i=1}^{n} (2^{\lambda_i} - 1),$$

where $\lambda_i \geq 0$. Alternatively expressed the condition states that $\alpha(d+n) \leq n$, where the α function counts the number of ones in the dyadic expansion of its argument. For example the Peterson 1-numbers are 1, 3, 7, $\ldots, 2^{\lambda_i} - 1, \ldots$. The first non-Peterson 2-number is 5, the first non-Peterson 6-number is 121. In [2] the following question is raised.

1.1 Peterson Conjecture. *As a module over the the Steenrod algebra, $F_2[x_1, \ldots, x_n]$ is generated by monomials whose degrees are Peterson n-numbers.*

In the case $n = 1$ for example it is easy to show that the collection of powers $x^{2^{\lambda_i}-1}$ is a minimal generating set. Indeed for any n, no monomial of the form

$$x_1^{2^{\lambda_1}-1} \cdots x_n^{2^{\lambda_n}-1}$$

can be a term of any polynomial in the image of the positively graded Steenrod algebra A_2^+ and these monomials must be included in any generating set. Clearly the total degree of a monomial of this type is a Peterson n-number. However for $n > 1$ the collection of such monomials is not itself a generating set. For example $x_1^2 x_2$ cannot be obtained this way.

A number of people have looked at the Peterson Conjecture, including Paul Selick and Eddy Campbell, who have proved it as far as $n = 5$. It turns out that a stronger statement is true [5] and we briefly recall its proof. From now on we talk about polynomials in the image of the Steenrod algebra when strictly speaking we mean in the image of the positively graded part of the algebra.

1.2 Theorem. *Let f be a monomial of degree d and suppose that e of its exponents are odd. If $\alpha(e+d) > e$ then f is in the image of the Steenrod algebra.*

The proof is based on a simple fact about the canonical map χ of the Steenrod algebra (Lemma 1 of [1]).

1.3 Lemma. *Let u, v denote polynomials in S and let θ be an element of A_2. Then*

$$u\theta(v) - \chi(\theta)(u)v$$

is in the image of the Steenrod algebra.

Now consider a monomial f with e odd exponents and write it in the form uv^2, where $u = x_1 \ldots x_e$ and v has degree t . Then

$$f = uSq^t v = \chi(Sq^t)(u)v \mod \operatorname{Im} A_2^+ .$$

The effect of any homogeneous element of the Steenrod algebra in degree t is zero on u if $\alpha(e+t) > e$. This demonstrates that f is in the image of A_2 if $\alpha(e+t) > e$. The proof of Theorem 1.2 is completed by observing that

$$\alpha(e+t) = \alpha(2e+2t) = \alpha(e+d) .$$

To obtain the Peterson Conjecture from Theorem 1.2 we let $n = e + r$ and observe that if $\alpha(d+e) \leq e$ then

$$\alpha(d+n) = \alpha(d+e+r) \leq \alpha(d+e) + \alpha(r) \leq e + r = n.$$

Hence if $\alpha(d+n) > n$ then $\alpha(d+e) > e$ and f is in the image of A_2^+ . This completes the proof. To see that Theorem 1.2 is stronger than the Peterson Conjecture note that $x_1x_2^2x_3^2$ is in the image of the Steenrod algebra because $\alpha(e+d) = \alpha(6) = 2 > 1 = e$, whereas $\alpha(n+d) = \alpha(8) = 1 < 3 = n$. A nice little example is $x_2^2x_3^3x_4^4x_5^5x_6^6$, which is in the image of the Steenrod algebra, whereas $x_1x_2^2x_3^3x_4^4x_5^5x_6^6$ is not.

2 Comments

If d is a Peterson n-number then we can always exhibit a monomial of degree d in n variables which is not in the image of the Steenrod algebra by taking each exponent to be one less than a power of two. There are of course other monomials in these degrees which are in the image of the Steenrod algebra and Theorem 1.2 provides a criterion for testing this. The test is not complete however. For example $x_1x_2^5$ is in the image but fails the test.

2.1 Problem. *Find a simple minded criterion on the exponents of a monomial which gives a necessary and sufficient condition for the monomial to be in the image of the Steenrod algebra.*

Ideally of course we would like such a criterion for arbitrary polynomials and a version of it for the odd prime case also.

Even as it stands however Theorem 1.2 is quite useful. The following statement is a sample result of what led Frank Peterson to his conjecture in the first place.

2.2 Theorem. *Let M^d be a closed smooth manifold such that all products of Stiefel-Whitney classes of length $> n$ vanish. Then either M^d is bordant to zero or $\alpha(d) \leq n$.*

The proof will appear in [3].

Another area where the Peterson Conjecture should be useful is in the work of Bill Singer [4]. Let

$$C = S/A_2^+ S$$

denote the cokernel of the the Steenrod algebra acting on the polynomial algebra. Then C is graded by degree and is a module over the semi-group ring $M(n)$ of $n \times n$ matrices over F_2 . Bill Singer constructs an iterated transfer map

$$\phi : E^2_{n,d} \longrightarrow C^{GL(n)}_{d-n}$$

from the E^2 term of the classical Adams spectral sequence to the invariants of C under the action of the general linear group $GL(n)$ over F_2 . He proves that ϕ is an isomorphism in cases $n = 1, 2$. In this context it is interesting to note the following result which was known to Frank Peterson in his initial work on the conjecture.

2.3 Proposition. *In the case $n = 2$ a minimal generating set for C consists of the equivalence classes of the monomials*

$$x_1^{2^{\lambda_1}-1} x_2^{2^{\lambda_2}-1}, \quad \lambda_1, \lambda_2 \geq 0,$$
$$x_1^{2^{\mu_1}-1} x_2^{2^{\mu_2}-2^{\mu_1-1}-1}, \quad \mu_2 > \mu_1 > 0\,.$$

In particular it is easy to check from these expressions that the dimension of C_d is never larger than three. For those people interested in modular representations of the general linear groups over F_2 it is worth noting that for a fixed n every simple $M(n)$ or $GL(n)$-module must occur as a composition factor of C_d for some d. The reason is that every simple module occurs in S and its first occurrence cannot be hit by a Steenrod operation from below since these operations are module maps. It is therefore of considerable interest to find composition series for the C_d, a more difficult task of course than locating just the trivial submodules of C_d as required in Singer's work, although this itself is hard enough. For $n > 1$ the dimension of the module S_d grows with d and the problem of finding a composition series for the whole polynomial ring is rather formidable. Experimental evidence so far supports the following encouraging possibility.

2.4 Conjecture. *For fixed n the dimension of C_d is bounded independently of d.*

For example in the three variable case a set of generators for C can be written down analogous to that of Proposition 2.3 and it is possible to verify conjecture 2.4 in this instance. An upper bound for the dimension of C_d is 27. In the two variable case the modules C_d can be described explicitly in terms of the trivial $GL(2)$-module D and the natural module N.

2.5 Proposition. *If $\alpha(d+2) > 2$ then $C_d = 0$. If $\alpha(d+2) = 1$ then $C_d = D$. When $\alpha(d+2) = 2$ there are two cases to consider according to whether the binary digits of $d+2$ are adjacent or not. In the first case $C_d = N$ and in the second case $C_d = N \oplus D$.*

More care is needed if we require the C_d as modules over the semigroup $M(2)$. In this case we interpret D as the determinant module, no longer trivial of course, and $N \otimes D$ has to be distinguished from N. The trivial $M(2)$-module appears in degree 0 but is not seen again. In the cases where C_d has dimension 3 there are short exact sequences which take one of the following forms

$$0 \longrightarrow N \longrightarrow C_d \longrightarrow D \longrightarrow 0,$$
$$0 \longrightarrow N \otimes D \longrightarrow C_d \longrightarrow D \longrightarrow 0.$$

These sequences do not split. Indeed C_d is monogenic as an $M(2)$-module. An instructive example is the case $d = 3$ where it can be checked that $x^3 + y^3 + xy^2$ generates a trivial $GL(2)$-module and xy^2 generates C_3 as an $M(2)$-module. Hence C_3 cannot split. Experience with examples like this leads to another question.

2.6 Question. *Are the $M(n)$-modules C_d monogenic for general n?*

The simple modules for $GL(3)$ are sufficiently accessible to put the decomposition of the C_d within reach but the details are rather complicated and combinatorial. We hope to report on this work elsewhere. For larger values of n we know less about the simple modules and the problems become very difficult. Indeed the hope is to exploit the Steenrod algebra in a significant way to say something interesting about the modular theory by using C_d as a better repository for the simple modules than the full symmetric algebra itself.

References

[1] S. Papastavridis: A formula for the obstruction to transversality. *Topology* 11 (1972), 415-416.

[2] F.P. Peterson: Generators of $H^*(RP^\infty \wedge RP^\infty)$ as a module over the Steenrod Algebra. *Abstracts Amer. Math. Soc.* 833-55-89, April, (1987).

[3] F.P. Peterson: A-generators for certain polynomial algebras. To appear.

[4] W.M. Singer: The iterated transfer in homological algebra. To appear.

[5] R.M.W. Wood: Steenrod Squares of polynomials and the Peterson Conjecture. To appear in *Math. Proc. Cambridge Philos. Soc.*

Printed in the United States
By Bookmasters